Climate Change and Intergenerational Justice

Synonymous with catastrophe and destructive tendencies, the Anthropocene provokes reflection on the limits of existing applications of ideas of responsibility, ecological agency and democratic justice. Youth campaigners, in particular, make emerging insights on the Anthropocene of central importance to an intersubjectively generated redefinition of the just society of the future. Given their span of affectedness, escalating rates of greenhouse gas emissions shape the ecological circumstances of generations to come and implicate them in harm relations they had no hand in creating. The realization is that human-inspired climate-destructive practices reverberate across plural time frames, thereby raising serious questions about the value of conventional interpretations of the copresence of sources of climate harm and their effects on the health and environmental living standards of all peoples. If injuries provoked by environmental degradation emerge across multiple time frames and affect generations differentially, where do we draw the boundaries of the just society, and how do we identify its most relevant subjects?

This book explores how such questions have ignited one of the most important debates on democratic justice in recent years – that between generations. For mobilized youth and future justice coalitions campaigning internationally, expanding resource inequalities (regionally and intergenerationally) are fundamentally issues of unfair exclusions and asymmetries in relations of power between generations. The book offers a comprehensive overview of new insights being generated through such debate on the limitations of democratic presentism, as well as current institutional applications of civil and human rights norms. It assesses overall how the metapolitical relevance of modernity's democratic project is being creatively redefined in terms more relevant to Anthropocene futures.

Tracey Skillington is Lecturer in Sociology in the School of Sociology & Philosophy, University College Cork. She is the author of *Climate Justice & Human Rights* (2017) and Guest Editor of a special issue of the *European Journal of Social Theory*, titled 'Perspectives on Climate Change' (8[3], 2015).

Sociological Futures

Series Editors:
Eileen Green, John Horne, Caroline Oliver, Louise Ryan

Sociological Futures aims to be a flagship series for new and innovative theories, methods and approaches to sociological issues and debates and 'the social' in the 21st century. This series of monographs and edited collections was inspired by the vibrant wealth of British Sociological Association (BSA) symposia on a wide variety of sociological themes. Edited by a team of experienced sociological researchers, and supported by the BSA, it covers a wide range of topics related to sociology and sociological research and will feature contemporary work that is theoretically and methodologically innovative, has local or global reach, as well as work that engages or reengages with classic debates in sociology bringing new perspectives to important and relevant topics.

The BSA is the professional association for sociologists and sociological research in the United Kingdom, with an extensive network of members, study groups and forums, and a dynamic programme of events. The Association engages with topics ranging from auto/biography to youth, climate change to violence against women, alcohol to sport, and Bourdieu to Weber. This book series represents the finest fruits of sociological enquiry, for a global audience, and offers a publication outlet for sociologists at all career and publishing stages, from well-established to emerging sociologists, BSA or non-BSA members, from all parts of the world.

Social Mobility for the 21st Century
Everyone a Winner?
Edited by Steph Lawler and Geoff Payne

Feeding Children Inside and Outside the Home
Critical Perspectives
Edited by Vicki Harman, Benedetta Cappellini and Charlotte Faircloth

Climate Change and Intergenerational Justice
Tracey Skillington

For more information about this series, please visit:
www.routledge.com/Sociological-Futures/book-series/SOCFUT

Climate Change and Intergenerational Justice

Tracey Skillington

Routledge
Taylor & Francis Group

LONDON AND NEW YORK

First published 2019
by Routledge
2 Park Square, Milton Park, Abingdon, Oxon OX14 4RN

and by Routledge
52 Vanderbilt Avenue, New York, NY 10017

First issued in paperback 2020

Routledge is an imprint of the Taylor & Francis Group, an informa business

British Library Cataloguing-in-Publication Data
A catalogue record for this book is available from the British Library

Library of Congress Cataloging-in-Publication Data
A catalog record for this book has been requested

ISBN 13: 978-0-367-66051-2 (pbk)
ISBN 13: 978-1-138-22297-7 (hbk)

Typeset in Times New Roman
by Apex CoVantage, LLC

For my mother, Angela

Contents

Figure

Acknowledgments

A special thank you to friends and colleagues for the rich and rewarding discussions on various themes addressed in this book: in particular, Patrick O'Mahony, Gerard Delanty, Piet Strydom, Hauke Brunkhorst, Richard Milner, Roddy Condon, Andrew Buchwalter, Graham Smith, Catherine Pearce and Jakob von Uexkull, Susana Batel, Andreas Niederberger, Jan Gehrmann, Ruben Langer and participants in the 'What's So Disturbing about Climate Change?' workshop at the Universitat Duisburg-Essen, Institut fur Philosophie in June 2016 for their critical input on theoretical arguments. Kind thanks also to postgraduate students in my Rethinking Borders seminar for their insightful comments and many hours of lively debate on these issues. Sincere gratitude to the editors of the Sociological Futures series – Eileen Green, John Horne, Caroline Oliver and Louise Ryan – for their continuous support. Finally, thanks to the College of Arts, Celtic Studies & Social Sciences in University College Cork for its support throughout this book project.

Introduction

Thinking differently about the future

Tracing the origins of the Anthropocene and the fossil fuel economy

Today, the life support systems of this planet are in serious decline. Both the stocks and flows of key elements, such as carbon, nitrogen, phosphorus and silicon, have been so badly affected by human-driven pollution activity that the Earth is no longer shaped by natural geological changes. Instead, scientists warn, the planet is moving rapidly into a biologically less diverse and climatically more dangerous state of existence (Steffen, Crutzen & McNeill, 2007). By far the most important factor contributing to these developments is the rise and ongoing expansion of the fossil fuel economy. Some 70 percent of global energy-related carbon dioxide emissions today are generated by fossil fuel production and consumption (IEA, 2018). This, however, is not a new phenomenon but, rather, one that has been emerging steadily for years. Neither is it a coincidence that the mass distribution of coal-burning technologies from the late eighteenth and early nineteenth centuries onward coemerged with a new geologic age known as the Anthropocene.[1] The Anthropocene defines a period in the planet's geological history when, for the first time, a peculiar breed of *Homo sapiens*, industrial capitalists, became primary agents of change, altering the Earth's surface, oceans, atmosphere and nutrient cycles to unparalleled degrees (Crutzen & Stoermer, 2000). With the further growth of fossil fuel-based energy systems and their associated technologies throughout the nineteenth and twentieth century (e.g., the invention of internal combustion engines), this pattern of destruction would continue to develop, triggering unprecedented ecological effects, including the threat of widespread ecosystem collapse.

To understand how this state of affairs emerged, it is necessary to look at its historical dynamic. In particular, the way specific actions and events would cumulatively allow what had been a fledging capitalist economy to achieve global supremacy. For instance, the decision to substitute firewood and charcoal as daily energy sources for coal from the eighteenth century on account of its abundance and easy extraction from near-surface sites. With the invention of the coal-burning rotary steam engine circa 1776 by James Watt (considered by many to be the quintessential tool of the Industrial Revolution), coal's commercial valuation as capitalism's first 'black gold' would be secured, fueling the rapid expansion of its new manufacturing industries and transportation technologies in ways that would transform society forever. Watt's steam engine linked coal to the inauguration of a new phase of global capitalist development but, also, a new phase in relations between humans and the carbon cycle (Crutzen, 2002: 23; Malm, 2016: 233). Coal was now needed in abundance to fuel the engines of capitalism's newly forged global interests, to transport resources from the continents, to dispatch its manufactured goods, and to secure the navigational supremacy of the colonial empires. So valuable was coal as an energy source during this period, its discovery often precipitated the colonial occupation of new territories. Commenting on the British occupation of Labuan from 1846 (an island off the north coast of Borneo), one newspaper described how the island 'appears to be one great coal field, for every large river intersects a coal-bed; and it seems only necessary to see and mineral is found'.[2]

Commitments to fossil-powered energy systems would continue to develop in the years that followed (e.g., the discovery of petroleum and gas energy sources, as well as the diesel-powered engine, etc.). By the mid-twentieth century, the harmful effects of mass fossil fuel consumption on the biophysical and biological systems of the planet (Steffen et al., 2011: 844–845) were beginning to show. The burning of fossil fuels would now give real definitional force to the destructive tendencies of the Anthropocene age and, simultaneously, the material basis of capitalism's violence against nature.

At no point in this development process was 'humankind' *the* agent responsible for inducing changes that would, in time, become the single largest threat to all planetary life. Rather, from the start, fossil-fuel-inspired geological changes were the product of the decisions and actions of a wealthy minority, and little, it seems, has changed in the years since. Just 100 companies worldwide today are responsible for 71 percent of total greenhouse gas (GHG) emissions (The Carbon Majors Report, 2017). When one considers that, globally, 1.2 billion people do not have access to electricity (International Energy Agency, 2017b), that the GHG emissions of one-sixth of the global human population are close to zero, or that the difference in modern energy consumption between a subsistence pastoralist in the Sahel and a high carbon consumer living in the West is thought to be larger by 1,000-fold (Smil, 2008: 259; Satterthwaite, 2009: 564), is it really appropriate to use species-wide descriptive categories to explain the 'causes' of climate change? Why, Malm and Hornborg (2014: 66) ask, does official policy discourse on climate change continue to speak of 'human inspired' climate change and consistently underscore how a qualitatively novel order of exploitation as the fossil fuel economy came into being or even why it continues to dominate climate change issues to this day? What we do know is that this economy was not created by chance. Neither does it continue to burn because it is indispensable to our lives. Rather, the primary agents promoting the fossil fuel agenda of today (e.g., the so-called oil shale revolution, deep-sea drilling projects, mining of coal through mountain top removal, etc.), as in the past, are amongst the wealthiest, most politically influential, and ecologically destructive. From its earliest years, the wealth and pollution generated by this economy were unevenly distributed and predicated on major divisions between peoples and power (capital accumulation, labor exploitation, class inequalities, colonial domination and widespread natural resource destruction). Inequality has, therefore, been a stable component of fossil-fuel-inspired climate change from the start. Yet the dominant viewpoint is that, historically, these developments were an essential part of humanity's progression towards more 'advanced' stages of civilization (see, for example, International Energy Agency 2017a). Rising concentrations of ozone in the troposphere, the pollution of rivers, oceans and so forth are framed as 'unfortunate side effects' of otherwise universally beneficial processes of industrialization and capitalist expansion. The more autonomous agency is accorded to climate change, the more passive symbolic formulations of its causes prevail (Fowler et al., 1979), and little attention is given to specific agents of harm – those, that is, whose largely unhindered operations have disproportionately contributed to the further advancement of the Anthropocene.

In the years since Crutzen and Stoermer (2000) first introduced their account of the Anthropocene age to the international community, the exploitation of energy from detrital carbon has continued to climb. In March 2015, scientists reported that monthly global average atmospheric concentrations of carbon dioxide exceeded 400 parts per million for the first time in 3 million years (since the mid-Pliocene era), a figure subsequently surpassed in September 2016 when concentrations increased to 403.3 parts per million (UN World Meteorological Association, 2016). With global energy use per capita projected to increase by another 12 percent by 2035 (BP Energy Outlook, 2035), the expectation is that concentrations of carbon dioxide will rise again (by roughly 40 percent according to the U.S. Energy Information Administration, see Scientific American, 2010). The impact of these changes, scientists argue, will be catastrophic, prompting further disruptions to average temperatures of air and water (Hamilton & Grinevald, 2015), as well as to the ecology of the world's oceans (which absorb an estimated 22 million tons of carbon dioxide every day [see NOAA, 2006]).[3] Today, all aspects of the Earth's natural features are being transformed by carbon-climate-actor interactions and their warming effects with a level of force that, potentially, is far greater than our abilities to control them.

As average global temperatures edge closer to a 1.5°C increase above preindustrial levels, lesser quantities of ecological goods and greater quantities of ecological harms are being transferred to newer generations. At the start of 2016, scientists reported that 2015 had been the hottest year in modern history, roughly 1°C warmer than the preindustrial average. That record was broken again when in early 2017 scientists concluded that average temperatures for 2016 had, in fact, been hotter still (NASA, 2017). Land and ocean temperature percentiles for January–December 2017 painted a similar picture, providing stark evidence that the 'climate tide is rising fast' (NOAA, 2018). The maize harvest failures in Africa left 6 million people on the brink of starvation in 2016 and, according to experts at the UK Met Office, are a worrying indication of how staple crop failures will become a more regular occurrence in the future, as temperatures rise and drought conditions deteriorate further (McKie, 2017). Already, the frequency of drought in dry subtropical regions has risen sharply (e.g., Cape Town's recent water crisis), affecting the most vulnerable sections of the population most (including children, the elderly and the disabled; see UNICEF, 2015). Similarly, extreme heat has fueled more regular wildfires across the United States, Central Europe and Australia. All evidence suggests that temperature extremes will prove increasingly difficult to avoid in the years ahead, placing more and more of the world's populations at the front line of climate change's devastating effects. The likely consequences of a 2°C rise in average global temperature will be serious, according to the Intergovernmental Panel on Climate Change Working Group III (2014b) and hard to control if dramatic changes are not introduced in the way we consume and produce energy. Many critical thresholds, however, have already been reached. Contamination of groundwater, oceans, food chains, and the atmosphere with toxic chemicals has reached dangerous levels of saturation (see National Geographic, 2009). The products of many years of high carbon living linger all around us in the atmosphere, the soil and oceans (potentially for millennia, according to scientists; see Frölicher, Winton & Sarmiento, 2014).

In a recent report to the UN General Assembly (February 2016), the Human Rights Council pointed to the fact that 'we are running out of time to avoid [climate change's] worst effects' (2016: 9, A/HRC/31/52). Unfolding pollution scenarios offer us a glimpse at a future of 'steering failure' when climate change no longer responds to efforts to control it (Klein, 2014). If attempts to reduce the amount of carbon dioxide and other greenhouse gases emitted into the atmosphere continue to fall short of what is required to reduce global warming, a 4°C rise in global mean surface temperatures will become a real possibility. At this point, the window of opportunity available to retrieve a functioning climate system will have passed (see IPCC, Climate Change, 2014b, Contribution of Working Group III). A key concern of the Paris negotiations in December 2015 was finding ways to prevent such temperature rises and steering the current energy system onto a safer path in the years ahead. Whether or not such an objective will be achieved remains to be seen. A 4°C rise in average temperature is certain to bring levels of danger we are entirely ill equipped to deal with (IPCC Working Group II, 2014b). At this temperature, tipping points will have been permanently passed and powerful feedbacks unleashed. Total yields of stable food crops, such as wheat, rice, and maize will decline steadily (International Scientific Congress on Climate Change, 2009; Intergovernmental Panel on Climate Change Working Group II, 2014b: 488), as will the amount of water stored in the soil. Extreme heat stress will reduce plant photosynthetic and transpiration efficiencies, negatively impacting root development and further exacerbating food insecurity (Intergovernmental Panel on Climate Change Working Group II report, 2014b: 796; Cropwatch, 2016). Chemical, physical and biological processes that modulate the functioning of the Earth will begin to break down and upward of 40 percent of species life will face extinction (see, for example, Barros et al., 2014). Such predictions have a definite 'nightmarish quality', as McKinnon (2012: 2) rightly observes. But if we are aware that global temperatures are being pushed beyond safe thresholds and that abrupt, unpredictable and potentially irreversible changes will follow, why are the real sociogenic sources of escalating climate change not being challenged?

CO_2 emissions from fossil fuel combustion and related industrial processes are and have always been the single largest contributor to global climate changes (Raupach & Canadell, 2010: 210). In spite of the growing magnitude of dangers posed by these pollutants, insufficient effort is being made to control them (e.g., the promotion of coal and nuclear energy as 'cleaner fossil fuels'; see Harvey's [2017] account of arguments presented by U.S. delegates at the Bonn climate talks, November 2017). Whilst all parties at this stage are clear as to the enormity of problems facing the international community and the necessity of reducing rates of CO_2 emissions, sustainable options are not being pursued with sufficient vigor or within a time frame adequate to offset the likelihood of future disaster. Rather than affirming commitment to a partnership model of resource justice and taking responsibility for harms already inflicted, the tendency has been towards the continuation of practices of 'dis-saving' (Rawls, 2001: 118). That is, depleting aggregate reserves of essential resources to levels insufficient to preserve safe living conditions into the future. How are we to explain the rational basis of these actions and a more general discounting of humanity's well-being as a lasting possibility (Nordhaus, 2007: 201–202)? The needs of the capitalist present, it would seem, have begun to take precedence over all other concerns. Traditional assumptions guiding the distribution of resources, most notably the importance of constraint, are openly compromised by the rising energy demands of a global economy (total energy consumption is expected to increase by a further 28 percent on present-day levels by 2040; see IEA, International Energy Outlook, 2017) or, at least, defined in terms more compatible with a narrow set of economic interests. Notions of fairness are conditioned by specifications as to which time frame and peoples matter most (i.e., the present). Principles of equity and sufficiency are overshadowed by market imperatives and the 'need' to take advantage of resources that are 'accessible', using all necessary means to secure their immediate availability (deep-sea drilling, hydraulic fracking, etc.).

For those critical of these arrangements, the truth of ongoing climate destruction must be defined in more specific rather than general terms (Loria, 2015) and issues avoided up to this point confronted more readily. For instance, if powerful economic and political interests fully back the continuation of a fossil fuel model of development for the future (e.g., see Institute for Policy Studies, 2013 on the World Bank's financing of new hydraulic fracking projects), how likely is a 40 percent reduction in global emissions by 2030 (Paris Agreement, 2015)? Is the stabilization of levels of carbon dioxide in the atmosphere really possible if fossil fuels still account for 77 percent of global energy in 2040, as the U.S. Energy Information Administration (EIA) confidently predicts?[4] Clearly, these are not reconcilable objectives, so how do we explain a nonrecognition of this fact? This book draws attention to the sociologically pertinent nature of these issues, in particular, how they evoke issues of exclusion, misrecognition, denial and inequality but, also, moments of critical reappraisal and a commitment to transformative action on the part of some. All will be explored in the context of deepening inequalities arising between generations in the distribution of the burdens of global climate adversities.

A sociological approach to intergenerational justice

While considerable attention has been devoted to the topic of intergenerational justice by political philosophy (e.g., Barry, 1999; Hiskes, 2009; Gosseries & Meyer, 2009; Tremmel, 2009) and to a lesser extent, political science (e.g., Sarat, 2014), in general, emphasis has been on how relations between present and future generations are governed by a series of duties that ought to be honored as a matter of right (Hiskes, 2009: 1). This is thought to be especially true of environmental relations and 'lifetime-transcending interests' (Thompson, 2009: 33). For instance, the provision of resources that make the flourishing of humanity, and indeed planetary life more generally, possible across time. This book seeks to bring a more distinctly sociological perspective to bear on these issues whilst acknowledging the value of various contributions to this discussion to date. Chiefly, its focus is on how intergenerational justice is interpreted by societal actors in response to 'live' ecological and related social, political and legal issues. It notes how the normative features of capitalist living, as well as their ecological and social costs are posed as subjects of fundamental dispute today. The societal value of this struggle ultimately shapes how and to what degree commitments to long-term environmental security can remake societal relations and change the cultural model by which we represent ourselves and act in a climate-challenged world (what Touraine [1988] refers to more generally as 'historicity' or the capacity of social actors to act upon society with a view to changing it).

Amongst the first cogent sociological collections to address climate change issues as inherently social and view their representation as intimately entangled with wider societal dimensions was that edited by Bronislaw Szersynski and John Urry (2010). Here contributors focus on the multiple social, political and cultural processes that sustain high carbon living over time, noting how these include not only routine social practices (e.g., see Shove, 2010) but also cultural viewpoints that prevent the exploitative logic of capitalism being subject to widespread critique (Beck, 2010). Urry examines how peak oil intersects with climate change to disrupt the notion of humanity's linear progression to better futures. In *What Is the Future?* Urry (2016) takes up many of these issues once again, paying particular attention to the performative dimensions of climate dystopias and the need to actively support the realization of more sustainable futures. How the project of democratization, even emancipation from 'catastrophic futures' (ibid.: 33–53) might be realized, however, is not explored by Urry. In *The Metamorphosis of the World*, Ulrich Beck (2016) notes how climate change, potentially, may trigger a societal 'metamorphosis' that, amongst other things, alters the way issues of justice and equality across territories, regions and generations are conceptualized and acted upon (e.g., see Beck 2016: 187–198). Beck, however, does not explore how the 'generations of the homo cosmopoliticus' (e.g., those with no memory of a world without the threat of climate change) actually respond to climate change concerns. Indeed, the political protests of this generation are dismissed by Beck 'not least because they are not unified by an idea of a better future' (Beck, 2016: 190), a viewpoint this book wishes to challenge. The analysis presented here sets out to examine how youth and fellow climate justice campaigners seek opportunities (especially those presented by law) to move society towards more sustainable outcomes and to redefine democracy as a partnership that is forged across generations. It notes the important contribution these actors make to collective processes of learning about the justice dimensions of climate change. More often than not, learning is generated by actors in dispute, as well as in communication with wider observing publics. Yet this inherently social aspect of climate justice has not been granted sufficient attention to date – a matter of some concern given the extent to which the societal significance of justice ideals (human rights and democratic principles) is radicalized by youth around the world today.

Intergenerational justice on climate change matters has become a project for real sociopolitical and legal change and must be acknowledged by sociological research as such. Youth emerge as societal carriers of alternative visions of climate justice, a development perhaps unforeseen by many. Traditionally, the tendency amongst social science communities has been to underestimate the contribution of youth to political discourse or policy debate on these and a series of other issues (Alderson, 2013: 7). The reality, however, is that youth 'are much involved in political campaigns, often with humour, costumes and elaborate posters against war, injustice, fracking' (Alderson, 2016: 135) and, more recently, against Donald Trump (e.g., orange 'Trump Baby' blimp, London protests, July 2018). As youth collaborations assert their voice across a range of concerns today, sociology is forced to take notice (see, for instance, Elliott and Earl [2018] who document an increase in levels of youth engagement in political campaigns in the United States, the majority of which occur outside of the formal political sphere; also see Crossley, 2008; Fletcher Fominaya, 2013; Roberts, 2015).

Youth are acutely aware of present environmental problems and the fact that they will be disproportionately affected by them in the future (those under the age of 24 constitute 41.6 percent of the current global population, see IndexMundi, 2018). Preceding youth mobilization on these issues are processes of critical reflection on and cognition of wrongdoing. It is the latter that enables a reconceptualization of prevailing societal 'structures, processes, social values and practices' (Jennings et al., 2006: 47) as a threat to future well-being. Yet such crucial stages of recognition, critical diagnosis and consensus building have not been granted sufficient sociological attention to date. Knowledge of what *ought to be* (e.g., justice across generations) only becomes societally transformative when it acquires 'capacity building' qualities. That is, when it begins to inform the social

actor's critical analysis of, reflection on and struggle against the visible and sometimes less visible action sequences and cognitive structures that perpetuate intergenerational inequalities and long-term ecological destruction.

If youth are not aware of structures of domination or of their own critical capacities to change them, there is little likelihood of justice principles empowering them
as 'the oppressed' (Freire, 1970). This book is really an exploration of how this
process of critical learning unfolds today amongst youth and future justice campaigners who challenge governments to take immediate action to address climate
change and safeguard the future. It considers how these actors attempt to build
stronger critical societal awareness of wrongdoing and, in the process, empower
themselves as agents of social change (Purdey et al., 1994). As a process of 'emancipation' from states of unknowing or denial of ecological, social and political
inequalities, the process of empowerment encompasses a crucial reassessment of
relations between generations as relations of domination and a commitment on
the part of mobilized youth (but other future justice campaigners also) to change
them. The following analysis is but a first step towards addressing these issues, as
they arise across differing national and international settings today. It considers
how sociological insights on power, exploitation and social agency might move
research on these issues forward and generate a deeper, empirically grounded interpretation of the ways ecological crisis triggers the processes of societal learning.

The aim is to advance a more critical account of how mobilized citizens, especially youth, interpret their role as custodians of sustainable futures and as members of 'inter-temporal democratic polities' (Bohman, 2014: 128) who challenge
various inequalities and search for the sociogenic roots of society's democratic
transformation (e.g., see Skillington, 2017: 231–260). Also, the book examines
how these actors see themselves as carriers of a different societal vision of climate
justice to that which systematically undervalues the needs of anything other than
the capitalist present. Youth insist that justice be defined in terms more relevant
to the 'deep time' of planetary well-being, that is, in a manner that sees our existence as intricately intertwined with multiple others located across space and time.
This means redefining relevant contexts for the application of principles of justice
and norms of democracy beyond just the here and now, incorporating a view of
the democratic community as a partnership across generations (e.g., see Intergenerational Foundation, 2018). In the chapters that follow, the critical normative
grounds of this argument will be explored in more depth, as will the part played
by specific actors in moving debates on climate justice forward legally, socially
and politically.

Arguably, one of the more interesting sociological aspects of these justice
campaigns is the extent to which their initiators assume that democratic change
is utterly realizable within the parameters of the existing legal framework and
pursue corrective action on that basis. Youth collaborate with international
legal experts to press for the further elaboration of certain shared institutional
expectations of justice (e.g., human rights commitments to health, development,
freedom of participation, as well as compliance with international agreements

on emissions reduction targets). Always, emphasis is placed on the violation of moral expectations of justice, in particular, the notion that youth are entitled to inherit a safe and healthy world and to challenge the current lack of symmetry in rights-based recognition. The assumption, consistently, is that law is capable of addressing their grievances. Not only those of the individual legal subject of the child (e.g., the rights of the child to 'survival', 'development' and to be heard in any judicial proceedings affecting their future, see UN Convention on the Rights of the Child, 1989) but, also generation-based injuries, that is, injuries generated through the knowing imposition of harm and the endangerment of the welfare of today's youth and generations to come through ongoing pollution practices. Such injuries are said to give rise to two types of inequalities. The first, synchronic inequalities, arise between generations living within the same society (in terms of the economic as much as environmental burdens generated by cumulative environmental problems, including lost agricultural production, damage to infrastructure by increased flooding and storms, rising costs of health, food or water crises, all borne disproportionately by youth). The second type, diachronic inequalities, arise more gradually across time on account of the temporal direction of accelerating pollution levels and their long-term effects on health and environmental living conditions.

   Because generation-based injuries arise slowly, youth insist that formulations of legal justice not be interpreted in a manner that is time specific or in a way that serves only the interests of those who exercise political and economic power in the present. 'The equal and inalienable rights of all members of the human family' (Universal Declaration of Human Rights, 1948) are said to transcend the temporal frame of the present and include all generations of humanity. A lack of knowledge of the specific identity of future peoples, youth argue, no longer provides sufficiently rational legal grounds for dismissing efforts to extend the principles of democratic justice to unspecified 'future others' (e.g., see Foundation for the Rights of Future Generations, 2018). Such rights interpretations may run contrary to the type of individualistic reasoning that still dominates traditional liberal interpretations of rights (where emphasis is placed on the living human subject), but there is a growing receptivity to such ideas in courtroom settings around the world (a development explored at length in Chapter 4 of this book). Sympathetic court judges acknowledge the constitutional validity of these actors' defense of the 'climate rights' of a 'non-identifiable group of persons needing protection' (i.e., future generations) or the 'general interests' of humanity, present and future.[5] In that, law acknowledges the contribution of youth to more contemporary explorations of how human rights law might better respond to the changing social, ecological and legal circumstances of justice today.

Ecological crisis and societal learning

For more than a decade, sociological perspectives on the cosmopolitan have challenged the discipline to adjust its analytical perspectives beyond borders of a social, cultural, legal and political kind and to account for processes of 'cosmopolitanization' occurring across territories, nationalities, regions, and cultures (e.g., Beck, 2006: 19). As a kind of involuntary or unforeseen transformation of everyday experiences and knowledge of the world, cosmopolitanization defines the course of all major sources of societal change today. In the case of climate change, the main stimulants to a decline in ecological circumstances do not apply differentially to some and not to others. All are exposed to the same forces of ecological destruction. Unavoidably, the global planetary sphere becomes the primary reference point when assessing not only the nature of the threat (Beck, 2002: 25–26) but also mechanisms needed to avoid future catastrophe. The understanding, therefore, is that an avoidance of danger requires a more globally sensitive perspective on climate change, one that is not limited to the present time frame or a purely nation-state outlook (Beck, 2009: 161). The cosmopolitanization of climate risks sparks an intense debate on where the boundaries of the just society ought to reside. For some, extending justice considerations to distant others (including future generations) is an indulgence we cannot afford given the present problems (e.g., growing water scarcity, drought, more frequent extreme weather events, etc.). For others, it is one we cannot avoid given the degree to which our endangerment is interconnected with that of others across time. Whatever our perspective on these issues, deepening climate problems force all to reflect more self-consciously on how we act upon the world and shape its future. For Beck, this has been one of the few advantages of cosmopolitanized risk. It compels a more 'cosmopolitan realist' approach to climate change (Beck, 2010: 10) and an acknowledgment at some fundamental level that the Westphalian order cannot deliver on promises of justice (for instance, its failure since international negotiations began to impose responsible limitations on major sources of ecological destruction). Indeed, justice procedures typically associated with this order seem more and more limited in their ability to address what are today transnationally situated problems. Ecological, social, political and territorial boundaries are no longer coterminous. A justice order that assumes that they are cannot succeed in the long term (Skillington, 2012: 1207).

A planetary frame of reference proves ever more difficult to avoid (Beck, 2016) as issues of pollution and resource scarcity become globally relevant. Climate change ushers in a cooperative imperative that runs far deeper than the politically constructed borders of the Westphalian sovereign state system can accommodate. Yet justice continues to be defined in largely state-centric terms. 'Normal justice' (Fraser, 2008) procedures are failing in the performance of basic regulatory functions, most notably, the protection of citizens from sources of serious harm (including the threat of ecological destruction, war, terrorism and displacement). However, the understanding also is that such failings do not arise solely from forces beyond states' control. In terms of the performance of democratic representational functions, problems arise. Most of those whose interests are deeply affected by climate change are excluded cohorts (e.g., populations under the voting age and generations not yet born). As the democratic deficits of 'normal justice' procedures grow more stark, the realization is that the current Anthropo-cene age is one of radical inequality. Major inconsistencies emerge between those who actively shape ecological futures and those who pay the ultimate price for current acts of ecological destruction. Such inequalities have a distinctly imperial-istic structure, according to Beck (2015: 1), on account of the extent to which the victims and perpetrators of climate harms do not overlap (generationally speak-ing). The span of affectedness of today's pollution practices reaches far beyond the present time frame (e.g., see Fang et al., 2013) and is predicted to have a det-rimental 'forcing effect' (NASA, 2017) on future climate conditions. As pollution levels continue on a steady upward trajectory, the expectation is that opportunities open to future generations to avoid ecological devastation will eventually disap-pear. Yet present misspendings of future peoples' environmental capital continue, and, as they do so, inequalities between generations expand.

To be able to enjoy their status as independent and free subjects, newer genera-tions must be guaranteed a sufficient threshold of resource availability (adequate supplies of essentials such as freshwater and stable food sources, arable lands, as well as a clean and safe atmosphere) as much as other types of capital (e.g., infra-structure, technology, material savings). Responsible levels of natural resource saving, however, are not guaranteed. Expectations of fairness and proportionality in the distribution of key resources are increasingly at odds with current prac-tices. A large-scale expenditure of nonrenewables, such as oil, coal, uranium and gas, make their long-term availability highly unlikely. Their depletion triggers detrimental ecological effects certain to reverberate for centuries (NASA, 2017). Youth come together around shared sentiments of frustration with governments regarding the ongoing mismanagement of long-term capital. Unfolding crises compel a critical diagnosis of the present as one of enduring system failure and political disappointment. Across regional and national borders, there is a synthe-sis of experiences of ecological degradation, employment insecurity, increasing poverty, debt and exclusion from decision making on issues that deeply affect the lives of all. As interpretive energies rise, inequalities lose their quasi natural-ness and are subject to more critical thematization. Anger is expressed at the way the decision to continue with fossil fuel energy pathways is made by a few and not by a democratic majority. Ongoing orientations towards institutional closure ensure that decision making on the best energy options for the future is not subject to broad public approval. The will to change harmful practices must, therefore, come from below, amongst those whose long-term safety is gravely jeopardized by poor regulatory arrangements. If current ecological practices can be shown to affect peoples living in the future most severely, then relations of justice can no longer be legitimately confined to the present. However hypothetical they may be at present, future generations are deeply relevant to current deliberations on justice (Skillington, 2015).

Increasingly, the demand is for a justice framework that responds to knowl-edge of anthropogenic climate destruction by extending traditions of democracy, freedom and right to a broader range of subjects. The realization is that prevailing models of justice are in need of revision, bound as they are by spatially and temporally limiting frameworks (Delanty & Mota, 2017: 20). This includes a revision of various ontological assumptions as to who is entitled to make claims to justice in this era of deepening ecological, social and economic problems (individuals, communities or generations whose interests and needs are threatened), to whom should justice considerations be extended (bounded political communities or wider transnational ones), and in what settings ought questions of justice be addressed (legal, political or social contexts)?

Claims to redistribution, for instance, traditionally understood in economic terms, are extended to a consideration of the distribution of the burdens of global climate change. New contingencies (e.g., disappearing territories), as much as enduring exclusions (e.g., communities of the developing world and future generations) call for a more open, cosmopolitan approach to distributive justice. A democratically 'enlightened anthropocentrism' (Chakrabarty, 2017: 41) is one that is sensitive to the globally relevant nature of these issues as they affect various cohorts (e.g., climate displaced populations, future generations, ecosystems) and contexts across time (endangered habitats, semiarid regions of the world, etc.). Youth insist that deliberations on these issues eventuate in more binding decisions that force major polluters to conform to democratically agreed-on emissions reductions targets and sustainable development measures for the future (e.g., see Earth Guardians, 2018).

In many ways, mobilization on these matters is driven by a greater consciousness of the present-ness of the dangers of ecological collapse, dangers that are no longer mere future trajectories (e.g., more regular and severe hurricane cycles, storm surges, prolonged drought and water shortages). The by now routine presence of ecological problems awakens a deeper sense of urgency in terms of the need for cooperative action and far-reaching institutional reform (Beck, 2003; Beck & Sznaider, 2006; Calhoun, 2010; Beck & van Loon, 2011). As pollution levels soar, 'really existing relations of interdependence' (Beck & Sznaider, 2006: 9) acquire a whole new degree of relevance. Suddenly, shifts in the carbon composition of the global atmosphere are of deep concern to all. Climate change and its adverse consequences, in being inescapable, require a level of global solidarity not witnessed before (e.g., an ethos of resource sharing in and across time rather than one of resource competition) and a distinctly cosmopolitan 'quality of mind' (Ossewaarde, 2007: 808) when addressing the evolving nature of relations between peoples and the rest of nature (Skillington, 2018). For others, it is more comprehensively a 'cosmopolitan imagination' (Delanty, 2006: 252, 2012; see also Delanty & Mota, 2017: 18) that is needed, one that envisages the growing interdependencies of all living systems on this planet as an opportunity to reconfigure the global relevance of sustainable development pathways. The fact that such pathways are not being pursued to any sufficient degree and that critical opportunities are tragically misspent is interpreted by youth as a blatant violation of their constitutionally grounded rights to a safe future.

Insights on the nature of social and geological worlds: the emancipatory potentials of the Anthropocene

Knowledge of the Anthropocene thus proves to have a strong societal significance over and beyond its power to define a new epoch in geological terms. Its advancement generates a greater desire for social, political and legal change but also, according to Chakrabarty (2009: 212), a new account of history. The desire is for a perspective on history that re-embeds nature-altering practices, especially those associated with industrial-capitalist developments, within the deeper time scales of the planetary system. In the process, the relationship between cause and effect is formulated in more expansive terms than the here and now. These changes have major implications for how society conceptualizes geological change and construes the relationship between past, present and future (Chernilo, 2017; Chakrabarty, 2015, 2017). Traditional assumptions regarding the relationship between presence (e.g., of victims of climate harms) and truth (proven acts of destruction), for instance, are subject to revision. The understanding now is that presence is no longer a perquisite for the imposition of harm (for instance, the discovery that greenhouse gases cause centuries of sea level rise and atmospheric pollution; see Chu, 2017). If the relationship between climate harm and effects can no longer be formulated as linear, with effects emerging as they do across multiple time frames, how are we to construct the boundaries of the just society and identify its most relevant subjects? This, in turn, raises a series of further questions on how we have traditionally conceptualized nature as largely passive in relation to the destruction imposed upon it (Chakrabarty, 2015: 204) and how we have conventionally formulated the justice dimensions of these practices (Chernilo, 2017). The Anthropocene necessitates a fundamental shift in perspectives on justice, as much as understandings of geological change. The work of Hornborg (2017) and Malm and Hornborg (2014) on the exploitative dimensions of the fossil fuel economy and its ongoing contribution to climate change makes this point abundantly clear (see, also, *The Lancet,* 2017).

As potentially catastrophic as the Anthropocene may prove to be, its advancement thus also provokes a moment of 'world disclosure' (Honneth, 2000: 116–127), when the micro foundations of global climate destruction are shown to be closely aligned with vested power interests and highly exploitative asymmetrical flows of cheap labor and natural resources to sites of concentrated power. Knowledge of the Anthropocene triggers societal learning potentials, as increasing numbers realize that the endurance of planetary life depends crucially on a collective capacity to address major sources of exploitation and harm. In this sense, the Anthropocene offers important insights on the nature of social, economic and political worlds, prompting new interpretations of 'right' and 'wrong relations', 'good' as well as 'bad' practices, and altering how the boundaries of the 'just society' are commonly drawn. Sociological research points to the importance of these developments (e.g., Strydom, 2015: 240), noting their contribution to a wider debate on how democratic traditions might be reinvigorated in response to a growing range of interrelated social, ecological and economic problems (e.g., see Chernilo, 2017). Attention shifts to those cultural and political components of modernity with the potential to address the calamitous forces that chiefly capitalist interests have unleashed onto the world. In particular, how ideals of equality, freedom and right might be better utilized in light of the fact that inequalities between peoples are expanding both regionally and intergenerationally (see Report of UN Secretary General, 'Intergenerational Solidarity and the Needs of Future Generations', 2013). The transformative capacities of the existing 'basic structure' (Rawls, 1999: 3–4) of justice are critically reassessed as citizens increasingly turn to law in the hope of initiating a more comprehensive program of intergenerational solidarity, sustainable development, as well as legal responsibility for environmental destruction. The realization is that justice must be redefined in deeper terms (temporally, spatially and intergenerationally) and current legal, political and social configurations subject to newer interpretations in response to rapidly changing circumstances.

The following chapters examine these processes in detail, noting how a growing awareness of the devastating impact of climate change contributes to this more critical dialogue on the need for action. Chapter 1 considers how a greater awareness of unprecedented levels of carbon pollution, resource scarcity, plastics in oceans, food and water sources, and so forth provokes a change in both the subjective experience of capitalist convenience living and the objective truth of its cumulatively destructive effects. Both, in turn, inform a critical diagnostic of the societal present as marked by inequalities that have been progressively expanding over time. Chapter 2 considers how these are interpreted specifically as intergenerational inequalities by future justice and youth campaigners. It notes how these actors challenge the settled convictions of prevailing short-term policy thinking and pressurize governments to extend principles of justice to newer subjects (e.g., future generations) and to problem areas exacerbated by declining climate conditions. Chapter 3 assesses how debate on these issues has opened up a new space of argumentation exploring the rights status of present youth and generations to come. It outlines some of the main arguments put forward in favor as well as against the development of a new normative framework of rights and responsibilities protecting the welfare of 'overlapping generations' of the Anthropocene. Chapter 4 documents the legal advocacy role played by youth coalitions across the world today, insisting upon a stronger institutional representation of their rights to a safe and democratic future. It looks at how claims to injury give real normative force to these actors' demands for accountability and a legal reform of existing regulatory arrangements governing the relationship between peoples and nature (Bernstein, 1992: 15–30). Chapter 5 accounts for those changes that have already been introduced in recognition of duties owed to future generations, while Chapter 6 considers prospects for the development of a more pluricentric approach to climate justice (encompassing a focus on the needs of the present and the future). Such an approach, it will argue, is entirely achievable within the parameters of the existing democratic legal order for two reasons chiefly. First is the extent to which both state constitutional and international legal systems remain open to new iterations of rights eligibility and subject relevance (for instance, the principle of openness embedded in various state constitutions, EU treaties and international law; see Brunkhorst, 2014: 455; Skillington, 2017: 246–247). The decades since 1945 have seen a progressive constitutionalization of international law and the further development of a two-tier governing structure consisting of elected government representatives and the 'peoples for whom these governments act' (Fassbender, 1998: 532). If the latter continue to interpret legal rights in terms more conducive to the needs of the planetary system as a 'generative commons' (Commons Abundance Network, 2018), then government will come under increasing pressure to adjust its justice perspective accordingly. The expectation is that 'cosmopolitanised states' will continue to respect the nation-transcending imperatives of universal law but at the same time remain primary implementers of new applications of these rights (Beck & Levy, 2013: 13).

Second, a pluricentric approach to climate justice is in keeping with the basic principles of intergenerational solidarity celebrated in the founding moments of the United Nations (e.g., UN Charter, 1945) and, in subsequent years, of the European Union (e.g., the Treaty on European Union, 1992).[6] The formation of both were witnessed by multiple generations who had experienced the horrors of war and persecution and now demanded the transformation of democracy both within and beyond the borders of their nation-states (Levy & Sznaider, 2006: 666). The understanding was that postwar and, in time, postcommunist generations would not only ensure that democracy continues into the future but that it would be organized in such a way that the present does not compromise the realization of basic liberties in the future (Bohman, 2014). Today this understanding of democracy (as intergenerational), as well as institutional commitments to its realization, continue to develop. Youth, in particular, have come to the fore of public debate exploring how the justice obligations of an explicitly intergenerational democratic institutional order can be made better to speak to the facts of deteriorating environmental conditions and endangered futures (e.g., by challenging practices of nonaccountability and closed-door policy). The critical attitude, cultivated historically through successive democratic revolutions and the Kantian mind-set of peoples who have fought repeatedly to liberate the future from oppression (Brunkhorst, 2014: 436), still encourages exclusions, inequalities and deprivations to feature strongly in public discourse on major challenges facing newer generations. It is precisely this critical attitude, as it is expressed in relation to the cross-generational impacts of deepening climate problems, that is the chief focus of this book.

Notes

1 At the end of August 2016, the International Union of Geological Sciences convened an international team of scientific experts to consider whether, in fact, humans have, through excessive pollution practices, pushed the planet into a third phase of the Quaternary period (2.6 million years to the present). Overwhelmingly, the opinion of committee members was that, yes, the age of the Holocene has ended and we have entered a

new phase of the Quaternary. According to atmospheric scientists Paul Crutzen and his colleague Eugene Stoermer (2000, International Geosphere-Biosphere Program), what distinguishes the 'Anthropocene age' is the progressively more central role played by humans in altering the planet's atmosphere, nutrient and carbon cycles (see also Raupach & Canadell, 2010). While the source of these changes are thought to be dispersed across centuries, the period of 'great acceleration' in the advancement of the Anthropocene is said to be the 1950s onward, when global populations began to rise steadily and a broader range of harmful pollutants were released into the environment, including radiation from coal-fired and nuclear energy production, the widespread use of nitrogen and phosphorous in fertilizers and the release of micro plastic particles into waterways and food chains. Not only has the discovery of the Anthropocene triggered a lively debate among geologists and environmental scientists, it has also provoked a dialogue among researchers in various social science and humanities fields as to the implications of these developments for society more generally.

2 See *Allen's Indian Mail*, 'Coal Fields of the Archipelago', 29 April 1856. Quoted in Malm (2016: 229).

3 Forty percent of carbon dioxide emitted remains in the atmosphere for 100 years, 20 percent for 1,000 years and the final 10 percent for 10,000 years (Union of Concerned Scientists, 2017).

4 See EIA's International Energy Outlook 2017. The expectation is that natural gas will remain the fastest growing fossil fuel in projections. Globally, natural gas consumption increases, on average, by 1.4 percent each year. 'Abundant natural gas resources', including supplies of 'tight gas, shale gas and coal bed methane', are expected to preserve 'the strong competitive position of natural gas' (EIA, International Energy Outlook 2017, Executive Summary: 1; also EIA, 2016) into the foreseeable future. Similarly, oil consumption has continued to rise steadily (from 3.8 billion tons in 1965 to 11.1 billion tons in 2007; see IEA, 2017b).

5 See ELAW (2015) Proceedings of the Hague District Court case, *Urgenda v. The State of the Netherlands* (June 2015), especially points 4.4 and 4.6, available at https://elaw.org/nl.urgenda.15 (accessed June 16, 2018).

6 See, for example, the Preamble of the UN Charter (1945) and its commitments to 'save succeeding generations from the scourge of war, which twice in our lifetime has brought untold sorrow to mankind'. The Treaty of the European Union (1992) similarly recalls 'the historic importance of the ending of the division of the European continent and the need to create a firm basis for the construction of future Europe'.

References

Alderson, Priscilla (2013) *Childhoods Real and Imagined: Volume 1: An Introduction to Critical Realism and Childhood Studies, Ontological Explorations*, London: Routledge.

Alderson, Priscilla (2016) *The Politics of Childhoods Real and Imagined: Practical Application of Critical Realism and Childhood Studies*, London: Routledge.

Barros, V. R. et al. (2014) *Climate Change 2014: Impacts, Adaptation, and Vulnerability. Part B: Regional Aspects, Contribution of Working Group II to the Fifth Assessment Report of the Intergovernmental Panel on Climate Change*. Cambridge: Cambridge University Press.

Barry, Brian (1999) 'Sustainability and intergenerational justice', in Dobson, Andrew (ed.) *Fairness and Futurity*, Oxford: Oxford University Press: 93–117.

Beck, Ulrich (2002) 'The cosmopolitan society and its enemies', *Theory, Culture & Society*, 19(1–2): 17–44.

Beck, Ulrich (2003) 'Toward a new critical theory with a cosmopolitan intent', *Constellations*, 10(4): 454–468.

Beck, Ulrich (2006) *Cosmopolitan Vision*, Cambridge, MA: Polity Press.

Beck, Ulrich (2009) *World at Risk*, Cambridge, MA: Polity Press.

Beck, Ulrich (2010) 'Climate for change, or how to create a green modernity?' *Theory Culture & Society*, Special Issue, *Changing Climates*, 27(2–3): 254–266.

Beck, Ulrich (2015) 'Emancipatory catastrophism: What does it mean to climate change and risk society?' *Current Sociology*, 63(1): 75–88.

Beck, Ulrich (2016) *The Metamorphosis of the World: How Climate Change is Transforming Our Concept of the World*, Cambridge, MA: Polity Press.

Beck, Ulrich & Levy, Daniel (2013) 'Cosmopolitanised nations: Re-imagining collectivity in world risk society', *Theory, Culture & Society*, 30(2): 3–31.

Beck, Ulrich & Sznaider, Nathan (2006) 'Unpacking the cosmopolitan for the social sciences: A research agenda', *British Journal of Sociology*, 57(1): 1–23.

Beck, Ulrich & van Loon, Joost (2011) 'Until the last ton of fossil fuel has burnt to ashes: Climate change, global inequalities and the dilemma of green politics', in Held, David, Theros, Marika & Fane-Hervey, Angus (eds.) *The Governance of Climate Change*, Cambridge, MA: Polity Press: 111–134.

Bernstein, Richard (1992) 'Philosophy, history, and critique', in Bernstein, Richard (ed.) *The New Constellation*, Cambridge, MA: MIT Press.

Bohman, James (2014) 'Children and the rights of citizens: Nondomination and intergenerational justice', in Sarat, Austin (ed.) *Intergenerational Justice*, New York: IDEBATE Press.

BP Energy Outlook (2035) available at https://www.bp.com/content/dam/bp/en/corporate/pdf/energy-economics/energy-outlook/bp-energy-outlook-2014.pdf (accessed: 5 July 2017).

Brunkhorst, Hauke (2014) *A Critical Theory of Legal Revolutions*, London: Bloomsbury.

Calhoun, Craig (2010) 'The idea of emergency: Humanitarian action and global (dis)order', in Didier, Fassin & Mariella, Pandolfi (eds.) *Contemporary States of Emergency: The Politics of Military and Humanitarian Interventions*, New York: Zone Books: 29–58.

The Carbon Majors Database (2017) 'CDP carbon majors report 2017' available at http://carbonmajors.org/ (accessed: 22 November 2017).

Chakrabarty, Dipesh (2009) 'The climate of history: Four theses', *Critical Inquiry*, 35(2): 197–222.

Chakrabarty, Dipesh (2015) 'The Anthropocene and the convergence of histories', in Hamilton, Clive, Bonneuil, Christophe & Gemenne, Francois (eds.) *The Anthropocene and the Global Environmental Crisis: Rethinking Modernity in a New Epoch*, Abingdon: Routledge.

Chakrabarty, Dipesh (2017) 'The future of the human sciences in the age of humans: A note', *European Journal of Social Theory*, 20(1): 39–43.

Chernilo, Daniel (2017) 'The question of the human in the Anthropocene debate', *European Journal of Social Theory*, 20(1): 44–60.

Chu, Jennifer (2017) 'Short-lived greenhouse gases cause centuries of sea-level rise', *MIT News*, 9 January, available at http://news.mit.edu/2017/short-lived-greenhouse-gases-cause-centuries-sea-level-rise-0109 (accessed: 22 July 2017).

Commons Abundance Network (2018) 'Commons as generative system' available at http://commonsabundance.net/docs/a-draft-for-action-items-of-the-people-sustainability-treaty-on-commons/ (accessed: 14 July 2018).

Cropwatch (2016) 'Impacts of extreme heat stress and increased soil temperature on plant growth and development' available at https://cropwatch.unl.edu/2016/impacts-extreme-heat-stress-and-increased-soil-temperature-plant-growth-and-development (accessed: 8 August 2016).

Crossley, Nick (2008) 'Social networks and student activism: On the politicizing effect of campus connections', *The Sociological Review*, 56(1): 18–38.

Crutzen, Paul J. (2002) 'Geology of mankind', *Nature*, 415(6867): 23.

Crutzen, Paul J. & Stoermer, Eugene F. (2000) 'The "Anthropocene"', *Global Change Newsletter*, 41: 17–18.

Delanty, Gerard (2006) 'The cosmopolitan imagination: Critical cosmopolitanism and social theory', *British Journal of Sociology*, 57(1): 25–47.

Delanty, Gerard (2012) 'The idea of critical cosmopolitanism', in *Handbook of Cosmopolitan Studies*, London: Routledge: 38–46.

Delanty, Gerard & Mota, Aurea (2017) 'Governing the Anthropocene: Agency, governance, knowledge', *European Journal of Social Theory*, 20(1): 9–38.

Earth Guardians (2018) 'Environmental lawsuits' available at www.earthguardians.org/youthvgov/ (accessed: 14 July 2018).

Elliott, Thomas & Earl, Jennifer (2018) 'Organizing the next generation: Youth engagement with activism inside and outside of organizations', *Social Media + Society* (January–March): 1–14.

Environmental Law Alliance Worldwide (ELAW) (2015) 'Proceedings of the Hague district court case', *Urgenda Foundation v. The State of the Netherlands* (June) available at https://elaw.org/nl.urgenda.15 (accessed: 16 June 2018).

Fang, Y., Mauzerall, D. L., Liu, J., Fiore, A. M. & Horowitz, L. W. (2013) 'Impacts of 21st century climate change on global air pollution-related premature mortality', *Climate Change*, 121(2): 239–253.

Fassbender, Bardo (1998) 'The United Nations charter as the constitution of the international community', *Columbia Journal of Transnational Law*, 36(3): 529–619.

Fletcher Fominaya, Cristina (2013) 'Youth participation in contemporary European social movements'. Report prepared for the partnership between the European Commission and the Council of Europe in the field of Youth available at https://pjp-eu.coe.int/documents/1017981/1668207/Youth_Participation_in_Contemporary_European_Social_Movements_Flesher_Fominaya_FinalPB_CFFPB_CFF_x2x.pdf/30c4dbcd-296e-411f-a338-ac033b3a98f6 (accessed: 14 July 2018).

Foundation for the Rights of Future Generations (2018) 'Seven building blocks for a future oriented democracy' available at https://generationengerechtigkeit.info/wp-content/uploads/2018/06/PP_Bausteine-f%C3%BCr-eine-zukunftsgerechte-Demokratie.pdf (accessed: 14 July 218).

Fowler, Roger, Hodge, Bob, Kress, Gunther & Trew, Tony (1979) *Language and Social Control*. London: Routledge.

Fraser, Nancy (2008) 'Abnormal justice' available at https://law.yale.edu/system/files/documents/pdf/Intellectual_Life/ltw_fraser.pdf (accessed: 16 July 2017).

Freire, Paulo (1970) *The Pedagogy of the Oppressed*, New York: Seabury Press.

Frölicher, Thomas Lukas, Winton, Michael & Sarmiento, Jorge Louis (2014) 'Continued global warming after CO_2 emissions stoppage', *Nature Climate Change*, 4: 40–44.

Gosseries, Axel & Meyer, Lukas H. (2009) (Ed.) *Intergenerational Justice*, Oxford: Oxford University Press.

Hamilton, Clive & Grinevald, Jacques (2015) 'Was the Anthropocene anticipated?' *The Anthropocene Review*, 2(1): 59–72.

Harvey, Fiona (2017) 'US switches focus of its Bonn event from clean energy to fossil fuels', *The Guardian* (10 November 2017) available at www.theguardian.com/environment/2017/nov/10/us-switches-focus-of-its-bonn-event-from-clean-energy-to-fossil-fuels (accessed: 22 November 2017).

Hiskes, Richard (2009) *The Human Right to a Green Future: International Rights and Intergenerational Justice*. Cambridge: Cambridge University Press.

Honneth, Axel (2000) 'The possibility of a disclosing critique of society: The dialectic of enlightenment in light of current debates in social criticism', *Constellations*, 7(1): 116–127.

Hornborg, Alf. (2017) 'Artifacts have consequences, not agency: Toward a critical theory of global environmental history', *European Journal of Social Theory*, 20(1): 1–16.

IndexMundi. World Demographics Profile (2018) available at www.indexmundi.com/world/demographics_profile.html (accessed: 16 June 2018).

Institute for Policy Studies (2013) 'World bank phasing out of coal, ramping up support for "Fracking"' available at www.ips-dc.org/world_bank_phasing_out_coal_ramping_up_support_for_fracking/ (accessed: 27 April 2016).

Intergenerational Foundation (2018) 'Do you feel it's fair?' available at www.if.org.uk/the-issue/ (accessed: 15 June 2018).

Intergovernmental Panel on Climate Change (2014a) 'Climate change 2014: Mitigation of climate change., Contribution of Working Group III to the Fifth Assessment Report of the Intergovernmental Panel on Climate Change' available at www.ipcc.ch/pdf/assessment-report/ar5/wg3/ipcc_wg3_ar5_frontmatter.pdf (accessed: 24 July 2017).

Intergovernmental Panel on Climate Change (2014b) 'Contribution of working group II to the fifth assessment report' available at www.ipcc.ch/pdf/assessment-report/ar5/wg2/ar5_wgII_spm_en.pdf (accessed: 17 June 2018).

International Energy Agency (2017a) 'Coal', available at www.iea.org/about/faqs/coal/ (accessed: 15 August 2017).

International Energy Agency (IEA) (2017b). 'Modern energy for all' available at www.worldenergyoutlook.org/resources/energydevelopment/ (accessed: 27 February 2017).

International Energy Agency (IEA) (2017c) 'International Energy Outlook 2017, Executive Summary' available at https://www.eia.gov/outlooks/ieo/pdf/exec_summ.pdf (accessed: 5 December 2018).

International Energy Agency (IEA) (2018) 'Global energy & CO_2 status report' available at www.iea.org/geco/ (accessed: 18 June 2018).

International Energy Outlook (2016) 'Coal-energy information administration', available at www.eia.gov/outlooks/ieo/coal.cfm (accessed: 19 May 2017).

International Energy Outlook (2017) 'International Energy Outlook 2017, *Executive Summary*' available at www.iea.org/Textbase/npsum/weo2017SUM.pdf (accessed: 22 June 2018).

International Scientific Congress on Climate Change (2009) available at https://sustainability.ku.dk/lectures-congresses/climatechangecongress09/ (accessed: 17 June 2018).

Jennings, Louise B., Parra-Medina, Deborah M., Hilfinger Messias, DeAnne K. & McLoughlin, Kerry (2006) 'Toward a critical social theory of youth empowerment', *Journal of Community Practice*, 14(1–2): 31–55.

Klein, Naomi (2014) *This Changes Everything: Capitalism vs. the Climate*. London: Allen Lane.

The Lancet (2017) 'Estimates and 25-year trends of the global burden of disease attributable to ambient air pollution: An analysis of data from the Global Burden of Diseases Study 2015' available at www.thelancet.com/journals/lancet/article/PIIS0140-6736(17)30505-6/fulltext (accessed: 20 November 2017).

Levy, Daniel & Sznaider, Natan (2006) 'Sovereignty transformed: A sociology of human rights', *British Journal of Sociology*, 57(4): 657–675.

Loria, Kevin (2015) 'One graphic shows exactly who is responsible for climate change' available at http://uk.businessinsider.com/who-is-responsible-for-climate-change-2015-7?r=US&IR=T (accessed: 15 August 2016).

Malm, Andreas (2016) 'Who lit this fire? Approaching the history of the fossil economy', *Critical Historical Studies* (Fall): 215–248.

Malm, Andreas & Hornborg, Alf (2014) 'The geology of mankind? A critique of the Anthropocene narrative', *The Anthropocene Review*, 1: 62, 69.

McKie, Robin (2017) 'Maize, rice and wheat: Alarm at rising climate risk to crops', *The Observer*, 16 July.

McKinnon, Catriona (2012) *Climate Change and Future Justice*, London: Routledge.

NASA (2017) 'Short-lived greenhouse gases cause centuries of sea level rise' available at http://news.mit.edu/2017/short-lived-greenhouse-gases-cause-centuries-sea-level-rise-0109 (accessed: 9 June 2018).

National Geographic (2009) 'Plastics breaks down in ocean, after all and fast' (20 August) available at https://news.nationalgeographic.com/news/2009/08/plastic-breaks-down-in-ocean-after-all-and-fast/ (accessed: 26 February 2017).

National Oceanic and Atmospheric Administration (NOAA) (2006) 'Carbon dioxide and our Ocean legacy' available at www.pmel.noaa.gov/pubs/PDF/feel2899/feel2899.pdf (accessed: 20 November 2017).

National Oceanic and Atmospheric Administration (NOAA) (2018) 'NOAA: 2017 was 3rd warmest year on record for the globe' available at https://www.noaa.gov/news/noaa-2017-was-3rd-warmest-year-on-record-for-globe (accessed: 6 June 2018).

Nordhaus, William (2007) 'Critical assumptions in the stern review on climate change', *Science*, 317: 201–202.

Ossewaarde, Marinus (2007) 'Sociology back to the publics', *Sociology*, 41(5): 799–812.

Purdey, A. F., Adhikari, G. B., Robinson, S. A. & Cox, P. W. (1994) 'Participatory health development in Rural Nepal: Clarifying the process of community empowerment', *Health Education Quarterly*, 21(3): 329–343.

Raupach, Michael R. & Canadell, Josep G. (2010) 'Carbon and the Anthropocene', *Current Opinion in Environmental Sustainability*, 2: 210–218.

Rawls, John (1999) *A Theory of Justice*, Harvard: Harvard University Press.

Rawls, John (2001) *Law of Peoples*, Harvard: Harvard University Press.

Roberts, Ken (2015) 'Youth mobilisations and political generations: Young activists in political change movements during and since the twentieth century', *Journal of Youth Studies*, 18(8): 950–966.

Sarat, Austin (2014) *Intergenerational Justice*, New York: IDEBATE Press.

Satterthwaite, David (2009) 'The implications of population growth and urbanization for climate change', *Environment & Urbanization*, 21: 545–567.

Scientific American (2010) 'Global emissions predicted to grow through 2035' available at https://www.scientificamerican.com/article/global-emissions-predicted-to-grow/ (accessed: 4 December 2018).

Shove, Elizabeth (2010) 'Social theory and climate change: Question often, sometimes and not yet asked', *Theory, Culture & Society*, Special Issue, *Changing Climates*, 27(2–3): 277–288.

Skillington, Tracey (2012) 'Climate change and the human rights challenge: Extending justice beyond the borders of the nation state', *International Journal of Human Rights*, 16(8): 1196–1212.

Skillington, Tracey (2015) 'Theorizing the Anthropocene', *European Journal of Social Theory*, 18(3): 229–235.

Skillington, Tracey (2017) 'Towards a transnational order of climate justice', in *Climate Justice and Human Rights*, Basingstoke: Palgrave Macmillan: 231–260.

Skillington, Tracey (2018) 'A deeper framework of cosmopolitan justice: addressing inequalities in the era of the anthropocene', in Delanty, Gerard (ed.) *Routledge International Handbook of Cosmopolitan Studies*, London: Routledge.

Smil, Vaclav (2008) *Energy in Nature and Society: General Energetics of Complex Systems*, Cambridge, MA: MIT Press.

Steffen, Will, Crutzen, Paul J. & McNeill, John R. (2007) 'The Anthropocene: Are humans now overwhelming the great forces of nature?' *Ambio*, 36(8): 614–621.

Steffen, Will, Persson, Asa, Deutsch, Lisa, Zalasiewicz, Jan, Williams, Mark, Richardson, Katherine, Crumley, Carole, Crutzen, Paul, Folke, Carl, Gordon, Line, Molina, Mario, Ramanathan, Veerabhadran, Rockstrom, Johan, Scheffer, Marten, Schellnhuber, Hans Joachim and Svedin, Uno (2011) 'The Anthropocene: From global change to planetary stewardship', *Ambio*, 40(7): 739–761.

Strydom, Piet (2015) 'Cognitive fluidity and climate change: A critical social-theoretical approach to the current challenge', *European Journal of Social Theory*, 18(3): 236–256.

Szersznski, Bronislaw & Urry, John (2010) 'Changing climates: Introduction', Special Issue, *Theory Culture & Society*, 27(2–3): 1–8.

Thompson, Janna (2009) 'Identity and obligation in a transgenerational polity', in Gosseries, Axel & Meyer, Lukas H. (eds.) *Intergenerational Justice*, Oxford: Oxford University Press: 25–49.

Touraine, Alain (1988) *Return of the Actor: Social Theory in Post-Industrial Society*, Minneapolis: University of Minnesota Press.

Treaty of the European Union (1992) available at https://europa.eu/european-union/sites/europaeu/files/docs/body/treaty_on_european_union_en.pdf (accessed: 13 July 2018).

Tremmel, Joerg Chet (2009) *A Theory of Intergenerational Justice*, New York: Routledge.

UNICEF (2015) 'Unless we act now: The impact of climate change on children' available at www.unicef.org/publications/index_86337.html (accessed: 31 March 2017).

Union of Concerned Scientists (2017) 'Why does CO_2 get most of the attention when there are so many other heat-trapping gases?' available at www.ucsusa.org/global_warming/science_and_impacts/science/CO2-and-global-warming-faq (accessed: 20 November 2017).

United Nations Charter (1945) available at www.un.org/en/charter-united-nations/ (accessed: 13 July 2018).

United Nations Convention on the Rights of the Child (1989) available at www.ohchr.org/en/professionalinterest/pages/crc.aspx (accessed: 15 June 2018).

United Nations Framework Convention on Climate Change (2015) 'Paris Agreement' available at https://unfccc.int/resource/docs/2015/cop21/eng/l09r01.pdf (accessed: 5 December 2018).

United Nations Human Rights Council (2016) 'Report of the special rapporteur on the issue of human rights obligations relating to the enjoyment of a safe, clean, healthy and sustainable environment' (A/HRC/31/52) available at www.ohchr.org/EN/HRBodies/HRC/.../A%20HRC%2031%2052_E.docx (accessed: 17 June 2018).

United Nations Universal Declaration of Human Rights (1948) available at http://www.un.org/en/universal-declaration-human-rights/index.html (accessed: 15 June 2018).

United Nations World Meteorological Association (2016) 'Carbon dioxide levels in atmosphere spike' available at https://public.wmo.int/en/media/news/carbon-dioxide-levels-atmosphere-spike (accessed: 15 June 2018).

Urry, John (2016) *What Is the Future?* Cambridge, MA: Polity Press.

Chapter 1

Relations between generations as relations of domination

Introduction

Large-scale violence against nature would appear in every sense to be a 'normal' component of the Anthropocene age. The utilization of nonhuman nature for energy, food, manufacturing and leisure is so central to our way of life and thinking that its nonpolitical status could not be more self-evident. Some of the oldest and most deep-seated ideas that have guided Western political thinking for centuries reflect this viewpoint: for instance, proclaiming capitalism's extractive instrumentalization of the Earth's riches a perfectly legitimate and noble endeavor (e.g., William Blackstone, 1765; Edmund Burke, 2000). The ongoing degradation of nature plays a constitutive role in the preservation of the current institutional order. At the core of its reasoning is the view that nature cannot be understood as injured or wronged so long as it not apprehended as normatively relevant (Krause, 2016). The entrenched nature of this viewpoint in everyday social life perhaps explains why our affective experiences of large-scale environmental destruction to date remain fundamentally directed towards the preservation of a development regime contrary to all attempts to connect us in a constitutive and nonviolent way to a wider ecology of matter. Instead, dominant political interpretations of the Anthropocene continue to reinforce our separateness from wider nature. Even now, as we stand at the tipping point of wholesale ecological disaster (global concentrations of carbon dioxide in the atmosphere are presently at a level not seen for more than 3 million years; see World Meteorological Organization, 2017), we remain inexplicably bound to a societal order that rationalizes the depletion of remaining resource reserves as 'the impersonal duty' of its calling (Weber, 1978). An 'implicatory denial' (Cohen, 2001: 7) of the dangers associated with this development order continues to hold much of contemporary policy discourse captive to a business-as-usual approach.[1]

In the period since the international community first began to commit to a series of legally binding agreements to reduce greenhouse gas emissions (1992), net carbon pollution has risen by more than 60 percent (see Le Quéré et al., 2014). What justificatory devices are used to render this situation as acceptable or the ongoing destruction of this planet's future somehow as 'necessary' in order to satisfy the energy demands of the present?[2] How can a capitalism of unlimited growth preserve the legitimacy of its operations in the face of irrefutable evidence of its practical irrationality? This chapter explores these issues in detail. It notes how a denial of the association between accelerating climate destruction and a fossil fuel capitalist economy may be evident at the level of public justification, but such denial does not characterize capitalism's understandings of the dangers that expanding environmental problems pose to its own resource interests. Responding to conditions of increasing scarcity worldwide, a major corporate project is under way to enclose the resource commons (Wallerstein, 2012). According to Christian Aid (2015), 203 million of hectares of land in the developing world have changed hands in the past decade alone, that is, a land area eight times the size of the UK. Harvey (2005) describes how this process unfolds as 'accumulation by dispossession' of wildlife of their habitats, of indigenous peoples of their ancestral lands, of communities the world over of essential resource supplies, and of future generations of their ecological inheritance. Natural resource justice is being denied to growing numbers as land acquisitions, as well as extractions, trade, and consumption of fossil fuels, metal ores, and nonmetallic minerals from all regions of the world rise exponentially (over the last four decades, global consumption of minerals rose by 22 billion tons to 70 billion; see UNEP International Resource Panel, 2016). Such dramatic increases in natural resource depletion inevitably contribute to deepening ecological problems but also to a more general crisis in the advancement of liberties beyond private property rights. The 'justness' of capitalism's resource acquisitions is guaranteed through an elaborate, legally sanctioned schema of property rights. The latter not only ensures that precious resource reserves can be bought, leased, or sold freely by those who possess contractual legal rights over them but, in doing so, ensure they are, in fact, not the subject of universal or free entitlement. Common rights to water, for instance, make little practical sense from a private property rights perspective.[3] In the absence of any serious attempt to impose responsible limitations on current large-scale acquisitions, the Anthropocene enters a qualitatively new phase of inequality. Rapid resource depletions are, therefore, more than forms of ecological destruction. They are also acts of legal violence (Skillington, 2015) against those denied the right to a fair portion of resources needed to secure a sustainable future.

As corporate actors dig deeper into remaining reserves of oil, gas, minerals and coal (and, with that, force carbon dioxide pollution levels to soar), it is difficult to imagine on what grounds such parties could claim these practices are just. Scientists estimate, on the basis of current rates of usage, that natural gas deposits will last for another 50 years and oil approximately 53 years. Similarly, the expectation is that our oceans will be fish-free by 2050 without a major change in fishing and pollution practices (UNEP 2010: 18–19). The crucial point, as Cohen (1985: 91) observes, is that the appropriation of a disproportionate amount of these and other essential resources on the part of some today not only opens up a greater possibility of exploitation of future generations' interests but, in fact, worsens the situation significantly for them. Increasingly, resource depletion practices fall foul of traditional liberal understandings of what a fair system of resource distribution requires (e.g., Locke's proviso that 'enough and as good' will be left for others or the Rawlsian principle of 'just savings'). The failure to conserve sufficient reserves in full awareness of the detrimental consequences of doing so is a reflection of the 'conscienceless reasoning' (Nelson, 1971: 169) of present-day capitalism, its need to free up resource capital 'as an antidote to excessive precaution' (Jasanoff, 2010: 242) and emphasize instead the importance of short-term enrichment. Indeed, this attitude goes a long way towards explaining why the 'distant suffering' (Boltanski, 1999) of future others does not register very strongly with contemporary reasoning or why scenarios such as sinking territories, scorched lands and soaring heat encourage, at most, a selective engagement with the 'grieveable' lives of others (Butler, 2004).

Or is the problem, as Marx reasoned, more general? Is society's failure to act responsibly in the interests of future others symptomatic of our alienation from 'species life' (Marx [1844], 1988: 162)? Perhaps it is, as German sociologist Rosa (2015) claims, indicative of the degree to which we have come to be conditioned by the cumulative demands of 'accelerated' capitalist living? A deliberate mystification of the underlining causes and consequences of climate change only seems to add to a sense of detachment from the suffering of distant others. Capitalist models of contemporary consumer living and their negative environmental and social effects exhibit a certain impenetrable quality that makes them appear immune to social transformation. With society seemingly incapable of generating anything other than short-term attentiveness to underlining problems, many wonder whether more sustainable worlds are even possible. In his essay 'Reification: A Recognition-Theoretical View', Honneth (2005) notes worrying tendencies towards a 'forgetfulness of recognition' of our common humanity and connection to the natural world. The 'pre-reflective' or affective realm of our existence, Honneth argues, is compromised by the project of rationalization, instrumental reasoning and the impersonal rules of capitalist living. Reification is said to define the essence of our affective relations with others, resulting in poor bonding experiences and a reduced capacity for empathy. Boltanski (1999: 43) explains how such tendencies encourage a preference to remain a spectator to the suffering of others and actively avoid being an introspector to the destruction occurring all around. The question of how largely avoidable tragedies (e.g., climate change, war, etc.) compel us to act is thereby avoided. Underlining Honneth's thesis on reification is the notion that the subject's freedom is actively curtailed. Domination in this instance is seen as arising from many influences embedded in the capitalist socialization process that the individual encounters on a daily basis. For Bourdieu, domination serves a more specific function: that is, to misguide publics in their assumptions about the world in which they live and, by doing so, preserve capitalist relations of inequality. Bourdieu's account of domination as 'misrecognition' offers a useful means of understanding why environmentally destructive practices are not challenged more readily by publics. However, such domination is not infallible. Occasionally, the lived experience of ecological harm informs the 'subjective truth' (Bourdieu, 2000: 202) of climate injustice and, in that, provides the actor with a basis for acting and exposing the false nature of political or corporate representations of the energy needs of society.

Capitalism's 'conscienceless reasoning' may seek to dampen belief in the possibility of alternative, less ecologically destructive worlds, but, in truth, it has never been entirely successful in eliminating all capacity to imagine the world otherwise (Fisher, 2009). Not all components of this society are completely absorbed by the materialities and calculating logic of its market-driven approach to resource management. Misrecognition of the objective truth of capitalism's wholesale destruction of ecological futures is by no means universally given. As a study of the mobilization efforts of youth and future justice coalitions across the world today, this books offers significant testimony in support of this view. Indeed, there is growing evidence to suggest that states of misrecognition of the truth of intergenerational inequalities are being challenged; what Bourdieu (2000; see also Burawoy, 2012) refers to as a 'para-doxa reversal' in states of unknowing.[4] 'Objective truths' regarding ongoing processes of ecological destruction are bridged with the 'subjective' (Bourdieu, 2000) and intersubjective truth of youth's endangerment. Epistemic, social and ethical sense making on climate change issues overlap more readily, giving rise to a scientifically informed yet situationally more grounded critical interpretive position (Jasanoff, 2010: 243).

At the heart of this process of 'cognitive liberation' (McAdam, 1988) is a critical reappraisal of sources of wrongdoing and the notion that cherished ideals of freedom and democratic right are being actively betrayed. Cognitive liberation is the process by which publics 'shift from one meaningful definition of reality to a new one, thus "making sense" of a situation, of facts and events, in new terms' (Donati, 1992: 155). New interpretative framings (Snow & Benford, 1988) of climate change as intergenerationally unjust emerge from a collective reappraisal of the 'taken for granted' or 'natural attitude' (Bourdieu, 1977) of present-day capitalist tendencies to discount the future and erase long-term resource supply as a topic of calculable concern. The ecological welfare and quality of life of today's youth and generations to come are now framed as being in serious jeopardy, as capitalism aggressively pursues remaining natural resource reserves and governments' enthusiastically support their efforts. Human history must be something more than just domination and ecological destruction (350.org, 2018). Rather, it must be conceived as a project of intergenerational solidarity (Brigstocke, 2016). In reaction to capitalism's logic of unlimited growth, mobilized youth call for greater constraint to be exercised over rates of consumption of essential resources. Indeed, constraint is defended as the most equitable response to scenarios of growing scarcity worldwide. A regulatory regime that only minimally registers this fact and consistently fails to protect long-term natural resource availability is said to be dysfunctional. The following sections examine how the exclusion of the question of constraint, and more generally intergenerational justice, from major policy considerations today is defended by dominant economic and political interests (through various strategies of denial). Second, it considers how this exclusion, as a form of metainjustice spread across many actions and decisions, inspires critical publics to reinterpret relations between generations as relations of domination and radical inequality. First, however, it will explore the peculiar nature of this domination, drawing on the insights of several theorists, including Pierre Bourdieu, Steven Lukes and Rainer Forst.

Accounting for relations of inequality that reach beyond the present

Traditionally, the tendency in sociological research has been to focus on relations of inequality as they arise between peoples living within the same time frame. Certain assumptions have dominated this research agenda. For instance, the importance of the physical presence of the subject of inequality. The problem with this sociological perspective, when applied to studies of climate injustice, is that it only partially explains the problem. Most of the burdens created by today's environmental practices will be borne by future peoples. Thus relations of inequality extend substantially beyond the present. Sociology is faced with the task today of having to account for inequalities that are neither wholly present nor absent from the current time frame (for instance, the long-term effects of accumulating greenhouse gases emissions in the atmosphere on drought conditions, ocean acidification and land fertility). Every so-called present point in environmental destructive practices is already one with the potential to reach far beyond the here and now (e.g., recent studies revealing how greenhouse gases will contribute to centuries of sea level rise; see MIT News, 2017), a characteristic that precludes inequalities ever being known or experienced entirely within the present historical moment. Relations of inequality in the distribution of climate change burdens, therefore, do not have a complete or fully decided character. The pollution practices of today are already shaping the ecological circumstances of future peoples. Similarly, the pollution generated by previous generations has had a 'forcing effect' on present climate conditions (see NASA, 2017). Not only do past environmentally destructive practices compromise the quantity of environmental goods transferred to newer generations but also the quality of essential resources available to them (interfering with such peoples' capacities to realize a future that is safe and plentiful). In this way, future peoples are implicated in harm relations they had no hand in creating. The consequences of damages generated today and in previous decades to health and well-being will be unavoidable in the future. Those most adversely affected by them will be unable to challenge decisions that allow such practices to continue. A research perspective that focuses only on inequalities arising between peoples in the present misses this additional sphere of injustice.

The intergenerational context is one where inequalities are expanding rapidly and therefore can no longer be ignored. The temporal direction of cumulative pollution levels produces serious asymmetries in power relations between generations (Gosseries, 2008). The insights of sociologist Stephen Lukes (2005) on the question of domination offer some preliminary guidelines as to how we might begin to extend established sociological reasoning on power relationships and explain how currently living polluting agents exert influence over future peoples in ways not always conducive to their interests. While the implicit focus of Lukes's (2005: 30) research is on parties that occupy the same time frame, his formulation of power as domination, nonetheless, offers a useful starting point for thinking about relations of inequality between generations, with A (e.g., present polluters) exercising power over B (future generations) in ways 'contrary to B's interests'. The most destructive forms of power, as Lukes (2005: 27) notes, are those used in ways that deliberately disadvantage vulnerable parties by limiting the range of options available to them (for instance, options available to control the likelihood of environmental disaster). Similarly, Pettit (2005: 92–93) describes agents as engaged in acts of domination when they exert an influence 'to the extent that they are in a position to interfere arbitrarily in the affairs' of others (see also Pettit, 2011). The knowing imposition of ecological harm through excessive pollution and the refusal to modify rates of resource depletion in response to wider changes in natural resource availability is clearly an example of exerting influence over others in ways that interfere with their long-term welfare prospects. In particular, it is an injustice against future generations who are unable to challenge such behaviors or to avoid their detrimental consequences.

By diminishing the quantity and indeed quality of available resources (e.g., essential food sources, such as fish stocks) in ways that threaten their long-term viability, present polluters (A) exert power over future generations (B) in a manner that is not conducive to their interests. According to the framework of power established by Lukes (2005), this form of power is, in effect, domination. Whilst pervasive in society, it must be distinguished from other varieties of societal power that are 'productive, transformative, authoritative and compatible with dignity' (Lukes, 2005: 109): for instance, decision-making units pursuing sustainable development as a collective goal or those attempting to push through changes in legislation governing energy options for the future (e.g., courts). The decision of many states to back new extreme energy exploration initiatives and to issue licenses to major energy companies to exploit remaining reserves of oil, gas and forests seem to run contrary to these more productive forms of power and, more generally, to the long-term interests of publics.[5] In principle, there is little in the way of barriers preventing states from pursuing alternative, low-carbon options and actualizing forms of institutionalized authority more compatible with principles of fairness and equity. A consistent failure to do so encourages many to question the ongoing legitimacy of states' role as trustees of a common ecological heritage. Why, critics ask, are less harmful alternatives not being pursued more readily with the interests of present and future generations firmly in mind?

When viewed in light of the current knowledge of the facts of environmental degradation, reassurances that 'containable' ecological risks will continue to be distributed across all communities as 'rough equals' prove increasingly unsatisfactory. No such relations of equality exist among world regions or generations.[6] The resource consumption choices being made today, in light of clear evidence of their long-term effects, are blatantly unjust. They represent forms of arbitrary interference (Pettit, 2005: 93) with the ecological fate of generations to come. Not only are viable, low-carbon resource conservation measures not being pursued with sufficient vigor, but the costs of not doing so are being raised considerably (a heightened threat of ecological disaster). Many scientists, for instance, believe the goal to limit global temperature rises to 1.5°C above preindustrial levels will be breached within a decade (see New Scientist, 2017). The failure of states to act to prevent global temperatures from soaring represents an act of domination, a denial of peoples' rights to a safe and sustainable future. To be able to secure functioning capability in the enjoyment of resources vital to safeguarding one's future, one must be in a position to gain access to them (i.e., sufficient food, freshwater reserves, as well as ecological conditions that promote health and well-being, including a safe atmosphere). The fact that the long-term availability of these resources is being actively undermined means that the functioning capability of billions to survive climate destruction is seriously threatened. All too apparent is the fact that this situation is the product of decision-making processes that neither generally nor reciprocally justify the risks they pose to those most affected by them. Critical publics draw attention to the democratically illegitimate nature of these arrangements and the problematic way governments continue to support the primary instigators of climate destruction.

Language, power and denial

At most, state political reasoning promotes a minimum cognitive registering of the facts of impending ecological disaster. But how is this practice of denial sustained when everywhere the evidence calls for a radical change in energy pathways? Clearly, it serves the interests of powerful economic elites to promote carbon-intensive energy options, but how is this practice framed as 'reasonable' and 'fair'? This question draws attention to what Fairclough (2010) describes as the representational function of language and the importance of its ideological properties to a preservation of certain states of unknowing (Ricoeur, 2004) the 'truth' of ecological harm doing.

What mediates contexts of experiencing, understanding, reasoning and communication about climate change risk is language. Not only does language facilitate a cognizance (or indeed denial) of climate change's plural realities, but as a fundamental component of communicative exchange on such issues, language is also a vital source of power (Forst, 2015: 115). The discursive political realm is where justifications for specific policy positions on fossil fuel depletion, for instance, are presented, debated, rejected or accepted. It is also the space where the cognitive basis of political power is vividly displayed. In its most elementary form, language is the primary medium used to motivate others to think or act in ways that they might not otherwise do (e.g., accept the 'need' to deplete nonrenewable resources at unsustainable rates). 'Noumenal power', Forst (2015: 116) adds, is the capacity to be socially effective in convincing others of the legitimacy of one's preferred choices (e.g., to prioritize the interests of some over that of others). Persuasion may emerge from an open critical debate amongst various parties where there is a reciprocal-general evaluation of several policy options on energy choices, for instance, and a decision on preferred ones arrived at democratically. Alternatively, persuasion may be achieved through a restriction of the range of 'viable alternatives' on the negotiating table, as well as the type of actors deliberating on these options. More often than not, compliance with preferred options (e.g., fossil fuel extraction) is secured by limiting the discursive space available in which the normative rightness of various energy pathways can be explored (i.e., whether *we ought to* be pursuing carbon-intensive energy choices in light of a growing body of evidence of their detrimental effects) and by whom.

In drawing attention to the importance of language to a delimitation of the cognitive space in which energy options are explored, the following analysis seeks to show how otherwise democratically motivated publics might be persuaded to accept a limited range of reasons why unsustainable energy options are best. Second, by deliberately limiting opportunities open to diverging perspectives on these issues (i.e., excluding those opposed to further fossil fuel extraction), carbon-intensive options are justified on the basis of a principle of lesser evil (Weizman, 2011). For instance, where the pursuit of further deep-sea gas and oil exploration is framed as a 'necessity', the best, 'cheapest' or 'most efficient' option available amongst a range of nonideal, ecologically damaging ones. Lesser evil reasoning is further supported by certain ideologically motivated interpretive strategies, such as that emphasizing the notion of abundance. The World Energy Resource Survey (2013: 7), for example, points to the 'enormity' of remaining coal, oil and natural gas reserves worldwide, while the International Energy Agency (2013: 3, 2016: 5) and the World Bank (2014: 2) both consistently offer reassurances that 'abundant low-cost supplies' of fossil fuels are sufficiently 'plentiful' to last for decades. The notion of abundance is promoted to advance the idea that the crisis precipitated by ongoing resource depletion may be temporarily averted and catastrophes predicted for the future confined to the category of 'distant possibilities'. On the basis of such evaluations, the 'gas market revolution' is said to have 'a healthy future'. Any ecological harm imposed is thought to be proportionate to the material gains that can be made. Clearly, this is a justice perspective focused on the interests of the present, one that strategically uses lesser evil reasoning to minimize the need for a reciprocal-general justification for the risks such projects pose to health and the environment more generally. No reference is made to the long-term effects of depleting remaining fossil fuel reserves (i.e., to health, ecosystems, the atmosphere, etc.). The possibility of a large-scale, globally relevant investment in alternative, renewable energy sources for the future is effectively closed down.[7]

As particular forms of institutional decision making and issue representation supported by relations of subordination and domination, 'noumenal power constellations' (Forst, 2017: 37) defend the continuation of high-carbon-energy options for the future. They do so chiefly by deliberately limiting opportunities available to discuss what is imaginable for globalized capitalist economies (e.g., the 'need' for further gas and oil exploration projects in the Arctic and other remote sites, oil sands and deep water drilling). On this basis, the resource demands of the global economy are presented as necessitating ever more extreme energy extraction measures. Energy options remain largely sealed off from any real accountability or critical debate about their costs to the environment and the future of the planet. In no way, critics argue, do these energy policy choices reflect the weight of free and open debate amongst diverging perspectives (Short et al., 2015; EarthJustice, 2017). On the basis of what democratic authority then do such decision-making practices continue to be supported? One fundamental source of authority in this regard is the unilateral right of states to control who gains access to resources on their territories and to decide how energy sources will be utilized. With preference being shown to the interests of living communities over future ones, the manner in which this authority is being asserted today has become problematic. Traditionally, the understanding has been that in exercising control over the distribution of scarce resources, sufficient supplies will be left for those who come after. Occupation of a specific territory by itself does not justify living generations' disproportionate exploitation of its essential resources to the exclusion of others. Consuming so much that subsequent arrivals have little or nothing, according to this Lockean formulation of the right to appropriation (see Locke, 1988 [1689] II ch v, para 33), is an injustice (Waldron, 2002). Constraint is seen as an essential response to conditions of natural resource scarcity in ways that current expressions of the state's role in the allocation of resource rights do not. Principles of reasonableness, fairness and proportionality are openly compromised as states succumb to the increasing resource demands of a capitalism 'without limits' (Heller, 1984: 46). For instance, moves to ensure sufficient water supplies are reserved to meet the water requirements of an expanding energy sector (expected to grow by as much as 85 percent by 2035 (IEA, World Energy Outlook, 2012; World Bank, 2013) even though over 700 million people currently do not have access to sufficient water (UN Water, 2018) and over 41 percent of the world's population live in river basins that are under serious water stress (World Wildlife Fund, 2018). With future climate dystopias now predicted to occur in the next decades rather than initial forecasts of a century (U.S. EPA, 2017), the Rawlsian assumption that objective conditions of resource scarcity will make a just cooperation in their distribution more likely (Rawls, 2001: 118) is sorely tested.

To counter the efforts of states and corporate actors alike to seal off the realm of justification for currently excessive rates of resource depletion from genuinely reciprocal practices of democratic justification, campaigners have sought ways of holding states accountable for breaches of legal obligations. Law has proven to be a particularly effective tool in this regard, enabling publics to challenge closed-door policies and achieve greater testimonial justice on environmental matters. Particularly prominent amongst those campaigning for justice on such matters today are youth coalitions. The latter represent the interests of a generation who may be described as the first with a truly socially situated perspective on the Anthropocene, those who can explain precisely how climate injustice is experienced differently on the basis of age and relationship to an endangered future. Youth see themselves as the primary victims of states' poor policy choices, the consequences of which (ecological but also financial costs of adaptation and cleanup costs following sudden extreme weather events) will have to be borne disproportionately by them in the future. The realization, therefore, is that the burdens of excessive pollution are distributed unfairly in geographical terms (with those residing in semiarid or low-lying regions of the world being most adversely affected by climate change) but also generationally. The majority of the life of youth today resides in a future under serious threat from climate change *and* poor efforts on the part of states to control it. Younger and older generations may occupy the same historical present, yet youth experience social time in qualitatively different terms (Sorokin & Merton, 1937: 615). One factor contributing to this difference is the absence today of the type of securities or opportunities that postwar generations took for granted (e.g., natural resource opportunities, economic, health, housing opportunities, etc.). Being similarly located in generational terms, youth across the world are not only threatened by ecological adversities but, also by institutional inertia and the failure on the part of governments to act at this crucial historical moment to secure more sustainable futures.

What Mannheim (1952: 297) describes as the 'stratification of experience' (*Erlebnisschichtung*) across different generations reveals itself today in how each cognitively and emotionally registers the injustice of climate change. As ecological, social and economic problems deepen, youth communicate increasing frustration with a political order that does not protect their interests and are drawn to those who voice their discontent with these arrangements. On account of their positioning within a highly stratified societal order of generations, youth encounter various forms of 'identity prejudice' (Fricker, 2007: 27), including a questioning of their credibility as legitimate participants in climate change policy negotiations. The more such prejudice is recognized by youth as unjust, the more states of misrecognition of the causes (contributing actors) and consequences of climate change are reversed. Building on an identity as excluded cohorts whose welfare and needs are systematically ignored by both states and capitalist interests, mobilized youth insist on a greater intersubjective recognition of their rights to a fair portion of resources essential for flourishing (allusions to constitutional rights, including every person's 'right to an environment that is conducive to health'), their right to participate in democratic decision-making processes affecting their future (e.g., Article 6[a] of the United Nations Framework Convention on Climate Change, 1992, and Article 12 of the Paris Agreement, 2015), as well as their legal right to be heard (e.g., Article 12 of the Convention on the Rights of the Child, 1989). The evidence would appear to suggest that youth demands for accountability and greater intergenerational equity are indeed intensifying (e.g., Nature and Youth, 2018; the Sink or Swim Campaign, 2018). However, these demands are not always taken seriously within the political domain, forcing many to turn to law in the hope of advancing a greater recognition of their rights status and countering efforts to deliberately diminish their credibility as legitimate sources of knowledge on climate injustice (what Fricker [2007: 1] refers to as 'testimonial injustice').

Youth as climate burden bearers

There is much evidence to suggest that youth, indeed, are emerging as primary losers in relation to climate change risks. The World Health Organization (2017) estimates that at present children in the developing world suffer more than 80 percent of the mortality and disease burden attributable to climate change (e.g., the expansion of vector-carried disease is expected to increase childhood mortality rates even further). In the case of children under the age of 5, this figure rises to 88 percent (see UNICEF UK, 2013: 5). Home to 1.9 billion children, or 85 percent of the world's current youth population, developing regions experience a disproportionate share of the extreme weather conditions attributable to climate change. Children in these regions will suffer heavily from heat stress, drought, crop failure and famine in the years ahead as climate conditions continue to deteriorate (see UNICEF, 2015: 1, 61; Currie & Deschenes, 2016; UNICEF, 2015: 6), and the risk of climate-linked diseases increases further. Rising average temperatures pose a major threat to the health of children on account of their underdeveloped thermoregulatory systems and smaller airways, which make them particularly susceptible to respiratory diseases (Zhiwei Xu, 2014). In the urban centers of more developed regions also, the incidence of asthma in children is rising steadily as emissions of nitrogen dioxide and sulfur dioxide, as well as ozone concentrations in the atmosphere grow (Sheffield et al., 2011). Worldwide, at present, 11–14 percent of children aged 5 years and older present asthma symptoms. An estimated 44 percent of these cases are linked to environmental problems, including air pollution, secondhand smoke and indoor pollution (see WHO, 2017). Beyond the immediate effects of increased particulate matter on the essential organs of the child, exposure to higher levels of ozone in the long term is also expected to increase the risk of early onset heart disease, stroke and cancer.

There are also the emotional and psychological impacts of climate change on children. Bourque and Willox (2014), for example, who examine the effects of extreme weather events and displacement on children, note a high rate of depression, anxiety and post-traumatic stress disorders. Similarly, Doherty and Clayton (2011) document a negative shift in the emotional well-being of children who are exposed to prolonged heat, drought, migrations and food and water scarcity in the developing world. Growing physical evidence of the destruction of ecosystems, rivers, air, wildlife and the like is triggering a type of psychological distress not officially recognized before now (see Albrecht et al. (2007); also Lancet Commission on Health and Climate, 2015: 1877; Kenyon, 2015). Albrecht and colleagues describe this psychological condition as 'solastalgia' and explain how it arises when our 'endemic sense of place is violated' by the pollution of land, rivers and seas, causing human distress levels to increase. Also referred to as 'ecosystem distress', solastalgia is exacerbated by the public's perceived sense of powerlessness or frustration with a lack of control over the unfolding destruction of their living natural environment (see Lancet Commission on Health and Climate, 2015). According to recent research conducted by UNICEF (2013), 75 percent of 11- to 16-year-olds interviewed in the UK expressed worry about how climate change will affect their future, as well as frustration with government's poor performance in tackling climate change. Psychologists are only beginning to understand how solastalgia gives rise to greater levels of anxiety and depression amongst youth (see, for instance, Kar et al., 2007). The loss of a sense of security and hope for a bright and safe future threatens the emotional, psychological and social development of our children.

For organizations such as Our Children's Trust (2016), these findings offer important new insights on changes in the health of the world's children. However, they also provide graphic evidence of how fundamental rights to the highest attainable standards of health (see Convention on the Rights of the Child, Article 24) are being violated. Across Europe and the United States, youth coalitions, together with organizations such as Our Children's Trust or Greenpeace and sympathetic climate scientists, are taking legal actions against states whom they believe are violating their constitutional right to health and a safe environment, as well as their fundamental right, as citizens, to be free of government actions that harm life and liberty (e.g., *Norwegian Government v. Youth and Nature, Greenpeace Norway*, October 18, 2016; complaint by *Kelsey Cascadia Rose Juliana, Xiuhtezcatl Tonatiuh M. Et Al. v. United States, Barack Obama et al.*, No. 6:15-cv-01517-TC, August, 2015). When political mechanisms of procedural justice fail to deliver on promises of equality of opportunity or participation in decision-making procedures, law is called upon to force state agencies to comply with international legal standards. In granting greater recognition to aggrieved parties' concerns, law provides an increasingly important setting for a consideration of whether carbon-intensive state policy choices are truly justifiable and in line with international standards of environmental protection.

Inspired by several recent legal victories, Client Earth, for instance, an international environmental activist law firm, brought the UK government to the Supreme Court recently, forcing it to comply with legal obligations to reduce air pollution and publish a report on how it plans to reduce pollution levels even further over the coming years (e.g., new carbon taxation proposals). While government ministers claimed they were not obligated to act on this issue as a matter of urgency, the Supreme Court disagreed and ordered the government to produce an air pollution plan as a priority. When it published a revised version of this plan in May 2017, campaigners rejected it on the grounds that it lacked specific proposals regarding carbon taxation or a diesel scrappage scheme. In this instance and in other legal settings across the world today, campaigners are asserting their right to challenge state actions. Law is explored as a mechanism of local empowerment and as a corrective to states' tendencies towards nonreciprocal justification for pursuing further, more extreme fossil fuel energy projects. Law also offers campaigners a platform with which to explore how universal ideals can be used as a means of exercising greater public authority over environmental concerns. As the number of legal cases brought by youth coalitions and climate justice campaigners around the world today continue to multiply, states find themselves under pressure to explain what have until now been nonreciprocal justifications for carbon-intensive models of development.

This, in turn, creates more opportunities for a wider discourse to occur on why alternative, more sustainable energy options are not being supported more readily by states. In this regard, courts play a crucial role, opening up new pathways of communication and accountability for the systematic failure of states to reduce emissions levels in line with international recommendations and to protect the interests of all citizens (e.g., constitutionally grounded duties to secure a safe, sustainable future and uphold the doctrine of public trust).[8] What emerges from a greater legal contestation of state duties in relation to environmental matters is a mix of contextually specific interpretations of rights to health, life, a safe environment, public participation and so forth, with more context-transcending endorsements of the right to have rights. In this way, the universal is 'endogenized' (Sassen, 2007: 436) inside national/local legal settings and simultaneously, externalized as new legal and political interpretations begin to spread across state borders and inspire further legal challenges elsewhere (e.g., prompted by similar legal actions in other countries, 25 Colombian youths filed the first climate change lawsuit in Latin America against their own government in January 2018), as well as a greater public communication on these issues.

Between justice ideals and lived realities: climate harms as violations of constitutional rights

Cumulatively, legal challenges contribute to a further discursive elaboration of what universal rights mean to a community at any given time (Butler, 1997: 366). Institutional interpretations of universal rights to development, health, life and self-determination, for instance, never fully exhaust what might be meant by these in the future. Ongoing revisions to historical interpretations of rights ensure that a component of the universal always remains open to new movements for democratic change (ibid.: 367), including those inspired by youth at present. Since 2011, youth have been working with legal experts to advance the Atmospheric Trust Litigation Campaign. This campaign seeks to hold states legally accountable for the destruction of a transnational public trust asset in urgent need of protection – the atmosphere. Harnessing the power of domestic courts and the doctrine of public trust, campaigners draw on established legal instruments, including universal rights, to advance 'a peoples-led approach to atmospheric justice'. The long-standing legal doctrine of public trust obliges states to protect certain resources held in common (e.g., the protection of wetlands, fish stocks, national parks, non-navigable waterways, coastlines etc.). Dating back originally to Ancient Roman Law, this doctrine has proven to be an indispensable legal tool to climate justice campaigners the world over who extend the scope of its relevance to a range of newer concerns, including deteriorating air quality and atmospheric pollution. With its emphasis on common ownership of basic natural resources, the doctrine of public trust is certainly not one that finds a great deal of support from the '*Homo oeconomicus*' (Weber, 1968) of late capitalist society whose attentiveness to the diminishing nature of essential resource reserves is primarily conditioned by the logic of private possession and less so by notions of common or transgenerational justice.

Even so, campaigners insist that, as temporary occupants of the Earth, we are no more than cotrustees of the atmospheric commons and are bound together not by private property contractual laws but by a framework of corollary and mutual responsibilities. States, they add, as legally authorized trustees, are obliged to protect the energy balance of the climate system for all by ensuring that CO_2 emissions levels are kept within safe limits and that essential, life-sustaining resources are preserved for the future. In the first week of May 2011, youth in cooperation with the nonprofit organization Our Children's Trust, put these demands into operation and initiated legal proceedings against several U.S. states. These and similar cases launched in the years since seek to accomplish through domestic litigation settings what has eluded international diplomatic treaty negotiations and emissions reduction targets to date, that is, an effective legal mechanism to force state agencies to respond more appropriately to GHG pollution and work towards achieving a greater climate balance that will benefit all into the future. Plans to launch a further series of legal actions are well under way. The most recent (April 2018) being a lawsuit filed by eight youths against the governor of the State of Florida for promoting, permitting and licensing commercial activities that exacerbate climate change and threaten citizens' constitutional rights to a safe environment. Elsewhere, including the Ukraine, Pakistan, India and Uganda, Our Children's Trust has partnered with local youth coalitions as well as transnational legal experts to launch legal cases against state agencies. In all instances, public trust principles and fundamental rights guide the main legal argument put forward. Both are contextualized in ways that local environmental circumstances and deteriorating global resource conditions demand.

Campaigners draw attention to glaring inconsistencies between official state commitments to climate change mitigation (e.g., UNFCCC, 1992; Paris Agreement, 2015) and the facticity of rapidly deteriorating environmental conditions. In that, campaigners raise interpretive dilemmas of immense value to society, stimulating processes of critical learning about the boundaries of the just society and the failure of states in the performance of environmental duties. At the center of this learning process is the understanding that the normativity of rights is never rooted entirely in their formal validity (i.e., the established legal order). Equally, it is generated by the interpretive efforts of an extra legal normative universe where publics continuously debate and reinterpret the validity of existing norms, creating in the process new varieties of rights meaning (Benhabib, 2010: 4). Current explorations of the rights of youth to inherit a healthy and sustainable world, for instance, give us some indication of how conceptualizations of rights are being repositioned today in response to rapidly changing ecological and social circumstances. Once again, rights are asserted as a constraint on an illegitimate use of power to further the interests of a few (current economic interests), but, in the process, they also enable new interpretations of justice to emerge (Derrida, 1990).

Being time locked or bound by the limits set by our birth and death does not, as these rights campaigners highlight, prevent living generations from contributing more positively to a justice framework that extends across time. As 'temporary stewards of the Earth's resources' (UN, 2013), present generations are duty bound to ensure that sufficient reserves of essential resources are conserved for the future, insofar as they are equipped with the capacity to do so. In this way, each living generation bequeaths to their successor a planet in at least as good a condition as that received from their predecessors (see also Brown Weiss, 1989). With each generation acting as a cotrustee of a resource commons that spans all generations, the understanding is that the Earth and its resources are not ours alone to exploit (see Sand, 2004: 57). Our duties are to manage the resource heritage of all of humankind rather than distribute it as private property amongst a few (ibid.: 6). To ensure that resources are indeed held safely in trust and that the rights of all peoples to a healthy environment are defended, the mobilization of legal protections is essential (EarthJustice, 2017). Human rights will be of little value to future generations if the latter's capacities to be self-determining continue to be actively undermined by excessive pollution and an exhaustion of those resources needed to sustain healthy living. The notion that each generation is a trustee of resources that are shared in common is a core element of future justice campaigners' thinking. It is also one at odds with the dominant model of resource distribution promoted by capitalist interests (i.e., private ownership). As disputes between these two opposing positions intensify, attention shifts once again to the legal responsibilities of states to ensure, in line with the doctrine of public trust, that resources such as 'running water, air, the shores of the sea' remain 'common to mankind' (The Institutes of Justinian, 2.1.1. (1867), quoted in Wood & Woodward, 2016: 648). Failure to fulfill such duties leads to notable anomalies in relations of ecological justice across generations. Chief executive of the international legal organization Client Earth, James Thornton, captures something of the mood of legal challengers at present, in particular their frustration with inept government policy on pollution control:

> The [UK] government . . . has dragged its heels for seven years while we choke on illegal and poisonous air; it's been hauled through the supreme and high courts and had to be forced by a judge to publish this delayed plan. We need to ask ourselves at what point will the government take this public health emergency seriously enough to start protecting our health and our Children's future, instead of the car companies?
>
> (Client Earth, James Thornton, quoted by McKie, 2017)

The reconstructive aim of these campaigners' work is twofold. First is to show how states, in failing to perform basic duties to protect citizens (e.g., through the subsidizing of further fossil fuel extraction), must be directed towards new development pathways for the future. Second is to demonstrate to publics more generally how society is still capable of being 'other' if it acts on its capacities to 'do other' and, in the process, 'become other' (i.e., more sustainable). A world warned of its vulnerabilities to ecological disaster cannot assume that those ideas of development and resource entitlement that have guided its thinking historically on a path to planetary destruction can continue to be the optimal ones for the future. Campaigners remind states of their political and legal obligations to protect resources held in trust. Failure to do so will diminish the power of future legislatures to promote the welfare of their people and the environment more generally. What is called for instead is a justice framework that works towards the realization of alternative 'social futures', that is, futures that are foreseen, planned, democratically coproduced and lived (see Urry, 2016). Cycles of ecological destruction must be broken and greater efforts made to substitute the transfer of environmentally bad products of our living to newer generations with renewable energy technologies, sufficient supplies of essential resources and functioning models of democratic organization. If intergenerational inequalities in life chances, development opportunities and environmental conditions are to be addressed more effectively, then justice must be defined in terms significantly broader than just the present. A core concern, therefore, is how the project of modernity will be used to initiate a more long-range approach?

When democratic decision-making arrangements are limited, the likelihood of achieving greater intergenerational justice is low (Bohman, 2005, 2011). At present, several obstacles must be addressed, including a lack of sufficient institutional support for more egalitarian regimes of resource sharing. A second and related problem is the absence of decision-making fora (both national and international) where the deeper justice implications of growing natural resource scarcity can be deliberated in an open and fair manner. Campaign groups have been instrumental in highlighting both of these issues and in pointing to the various ways the democratic freedoms of peoples to realize greater intergenerational justice are systematically denied (Skillington, 2017). The realization, increasingly, is that the threat of climate disaster derives from two main sources: expanding pollution practices and ongoing institutional closure to alternative models of sustainable living. Campaigners accuse governments of complicity with corporate actors' 'criminal acts of ecological destruction', that is, acts of pollution that seriously compromise the health of human populations, species and ecosystems, both present and future (see the World Future Council, 2013). Major crimes are said to include Arctic drilling, bottom trawling and razing the rainforests. More than 'harms', such acts of ecological destruction are seen as crimes on account of the traceable connections that can be established between intent (to pollute) and multiple adverse environmental and social effects. Campaign groups call attention to a growing body of scientific evidence of traceable harms (e.g., the CDP Carbon

Majors Report [2017] pinpoints 41 public investor-owned companies, 16 private investor–owned companies, 36 state-owned companies and seven state producers as 'responsible' for 71 percent of global carbon emissions) and demand that more effective procedures of accountability and democratic justification be introduced.

As an inescapably cosmopolitizing force, climate change compels a more serious discussion on how settled justice convictions might be better used to address its major challenges. As it does so, the more 'abstract' dimensions of anthropogenic emissions of CO_2 are translated into the culturally familiar, politically relevant and existentially real. The following chapter examines the type of critical evaluations that future justice and youth campaigners bring to bear on this translation process, noting, in particular, how these actors frame climate change as the product of dysfunctional democracies and 'abnormal' justice relations (Fraser, 2008). It also notes these actors' efforts to reignite the potentials of democratic societies to transform relations of inequality and insist that institutional resources be used more constructively to reform decision-making arrangements and strengthen public reasoning on the real implications of uncontrolled climate change.

Notes

1 The recommendations of expert bodies, such as those of the Lancet Commission on Health and Climate Change (2015: 1883), is that coal be immediately phased out of the global energy mix in the interests of long-term health and well-being.
2 Energy use worldwide increased 54 percent between 1990 and 2013 (World Bank, 2016).
3 For example, the claim of the former chairperson of the global food giant, Nestle, that 'access to water is not a public right' nor a human right (see Samson, 2016).
4 In *The Logic of Practice* (1997: 110), Bourdieu explains how doxa is the 'unanimity effect' in social collectivities, or shared commitment to the 'presuppositions' of 'the game' (ibid.: 66), but, ultimately, doxa is 'an act of misrecognition, implying the most absolute form of recognition of the [established] social order' (Bourdieu, 1984: 471). In *Pascalian Meditations* (2000: 130–131), Bourdieu further clarifies how doxa defines the socially dominated nature of the natural attitude, the sense of knowing one's place in the world expressed in statements such as 'that's not for the likes of us' (Bourdieu, 2000: 185).
5 In May 2017, the U.S. Department of Interior sold a record-breaking $2.6 billion in development bids throughout the Chukchi Sea, above the Bering Strait, to gas and oil exploration companies.
6 While all regions eventually will experience the physical effects of deteriorating climate conditions, poorer countries who have contributed least to this problem will suffer its effects the most. Developed countries, with larger, more diversified economies, who historically accumulated their wealth through fossil fuels, will fare better (see Oxfam Ireland, 2018).
7 Instead, energy options, such as nuclear energy, are promoted by the World Nuclear Association (June 2016) as 'the most environmentally benign way of producing electricity'.
8 Under the public trust doctrine, no government can legitimately relinquish its obligations to protect resources held in trust without simultaneously diminishing the power of future legislatures to promote the welfare of the people. Resources vital to the flourishing of future generations are threatened by climate change. The failure of current government to protect the climate system is, therefore, a violation of the doctrine of public trust.

References

350.org (2018) 'We believe in climate justice' available at https://350.org/about (accessed: 15 July 2018).

Albrecht, Glenn, Sartore, Gina, Connor, Linda, Higginbotham, Nick & Freeman, Sonia, Kelly, Brian & Stain, Helen (2007) 'Solastalgia: The distress caused by environmental change', *Australasian Psychiatry*, 15, Supplement 1(1): S95–S98.

Benhabib, Seyla (2010) 'Human rights, sovereignty and democratic iterations', *Our Common Future*, Hannover/Essen (4 November).

Blackstone, William (1765–1769) *Commentaries on the Laws of England*, Oxford: Clarendon Press.

Bohman, James (2005) 'The democratic minimum: Is democracy a means to global justice?' *Ethics & International Affairs*, 19(1): 101–116.

Bohman, James (2011) 'Children and the rights of citizens: Non domination and intergenerational justice', in Sarat, Austin (ed.) *Intergenerational Justice*, New York: IDEBATE Press: 128–141.

Boltanski, Luc (1999) *Distant Suffering: Morality, Media and Politics*, Cambridge: Cambridge University Press.

Bourdieu, Pierre (1977) *Outline of a Theory of Practice*, Cambridge: Cambridge University Press.

Bourdieu, Pierre (1984) *Distinction*, London: Routledge.

Bourdieu, Pierre (1997) *The Logic of Practice*, Cambridge, MA: Polity Press.

Bourdieu, Pierre (2000) *Pascalian Meditations*, Cambridge, MA: Polity Press.

Bourque, François & Willox, Ashlee Cunsolo (2014) 'Climate change: The next challenge for public mental health', *International Review of Psychiatry*, 26(4): 415–422.

Brigstocke, Julian (2016) 'Exhausted futures', *GeoHumanities*, 2(1): 92–101.

Brown Weiss, Edith (1989) *In Fairness to Future Generations: International Law, Common Patrimony and Intergenerational Equity*, Tokyo: United Nations University.

Burawoy, Michael (2012) 'The roots of domination: Beyond Bourdieu and Gramsci', *Sociology*, 46(2): 187–206.

Burke, Edmund (2000) *On Empire, Liberty and Reform: Speeches and Letters*, ed. David Bromwich, New Haven, CT: Yale University Press.

Butler, Judith (1997) 'Sovereign performatives in the contemporary scene of utterance', *Critical Inquiry*, 23(2) (Winter Edition): 350–377.

Butler, Judith (2004) *Precarious Life: The Powers of Mourning and Violence*, London: Verso.

The Carbon Majors Database (2017) 'CDP carbon majors report 2017', July, available at http://carbonmajors.org/ (accessed: 16 November 2017).

Christian Aid Ireland (2015) 'Large-scale land acquisitions', available at https://www.christianaid.ie/sites/default/files/2018-02/land-matters-large-scale-acquisitions.pdf (accessed: 5 December 2018).

Cohen, Gerald A. (1985) 'Nozick on appropriation', *New Left Review*, 1(150) (March–April): 89–105.

Cohen, Stanley (2001) *States of Denial: Knowing about Atrocities and Suffering*, Cambridge, MA: Polity Press.

Currie, Janet & Deschênes, Livier (2016) 'Children and climate change: Introducing the issue', *The Future of Children: Children and Climate Change*, 26(1) (Spring): 3–4.

Derrida, Jacques (1990) 'The force of law: The mystical foundations of authority', *Cardozo Law Review*, 11: 920–1046.

Doherty, Thomas J. & Clayton, Susan (2011) 'The psychological impacts of global climate change', *American Psychologist*, 66(4): 265–276.

Donati, Paolo (1992) Political discourse analysis, in Diani, Mario and Ron Eyermann (eds.) *Studying Collective Action*, London: Sage:136–167.

EarthJustice (2017) 'Out of control out west' available at https://earthjustice.org/features/extreme-energy-out-of-control-out-west (accessed: 16 November 2017).

Fairclough, Norman (2010) *Critical Discourse Analysis: The Critical Study of Language*, Harlow: Longman.

Fisher, Mark (2009) *Capitalist Realism*, Ropley: Zero Books.

Forst, Rainer (2015) 'Noumenal power', *The Journal of Political Philosophy*, 23(2): 111–127.

Forst, Rainer (2017) *Normativity and Power: Analyzing Social Orders of Justification*, Oxford: University of Oxford Press.

Fraser, Nancy (2008) 'Abnormal justice', *Critical Inquiry*, 34(3): 393–422.

Fricker, Miranda (2007) *Testimonial Injustice: Power and the Ethics of Knowing*, Oxford: Oxford University Press.

Gosseries, Axel (2008) 'On future generations' future rights', *The Journal of Political Philosophy*, 16(4): 446–474.

Harvey, David (2005) *The New Imperialism*, Oxford: Oxford University Press.

Heller, Agnes (1984) 'Marx and modernity', *Thesis Eleven*, 8(1): 44–59.

Honneth, Axel (2005) 'Reification: A recognition-theoretical view', *The Tanner Lectures on Human Values*, University of California Press, Berkeley (14–16 March) available at http://tannerlectures.utah.edu/_documents/a-to-z/h/Honneth_2006.pdf (accessed: 16 May 2017).

International Energy Agency (2012) 'World energy outlook 2012' available at http://webcache.googleusercontent.com/search?q=cache:su3aGY6uDMgJ:www.iea.org/Textbase/npsum/weo2012sum.pdf+&cd=1&hl=en&ct=clnk&gl=ie (accessed: 1 June 2018).

International Energy Agency (2013) 'Resources to reserves: Oil, gas and coal technologies for the energy markets of the future' available at www.iea.org/publications/freepublications/publication/Resources2013.pdf (accessed: 17 August 2017).

International Energy Agency (2016) 'International Energy Outlook 2016' available at https://www.eia.gov/outlooks/ieo/pdf/0484(2016).pdf (accessed: 5 December 2018).

Jasanoff, Sheila (2010) 'A new climate for society', *Theory, Culture & Society*, 27(2–3): 233–253.

Kar, Nilamadhab, Mohopatra, Prasanta K., Nayak, Kailash C., Pattanaik, Pratiti, Swain, Sarada P. & Kar, Harish C. (2007) 'Post-traumatic stress disorder in children and adolescents one year after a super-cyclone in Orissa, India: Exploring cross-cultural validity and vulnerability factors', *BMC Psychiatry*, 7: 8.

Kenyon, Georgina (2015) 'Have you ever felt Solastalgia?' *BBC Future*, 2 November.

Krause, Sharon (2016) 'Politics beyond persons: Political theory and the non-human', *Political Theory* (June Edition): 1–13.

Lancet Commission on Health and Climate (2015) 'Health and climate change: Policy responses to protect public health', 7 November, available at www.thelancet.com/pdfs/journals/lancet/PIIS0140-6736(15)60854-6.pdf (accessed: 13 June 2017).

Le Quéré, Corinne et al. (2014) 'Global carbon budget 2014', *Earth Systems Science Data Discussions*, 7: 521.

Locke, John (1988) *Two Treaties of Government*, Cambridge: Cambridge University Press.

Lukes, Steven (2005) *Power: A Radical View*, Second Edition, Houndmills: Palgrave Macmillan.

Mannheim, Karl (1952) 'The problem of generations', in Kecskemeti, Paul (ed.) *Karl Mannheim: Essays*, London: Routledge.

Marx, Karl (1988) *Economic and Philosophic Manuscripts of 1844*, New York: Prometheus Books.

McAdam, Doug (1988) 'Micromobilization contexts and recruitment to activism', in Klandermans, Bert, Kriesi, Hanspeter & Tarrow, Sidney (eds.) *From Structure to Action*, Greenwich, CT: JAI Press: 125–154.

McKie, Robin (2017) '"Appalling" air pollution plan faces fresh court battle', *The Observer*, 7 May.

MIT News (2017) 'Short-lived greenhouse gases cause centuries of sea level rise' available at http://news.mit.edu/2017/short-lived-greenhouse-gases-cause-centuries-sea-level-rise-0109 (accessed: 27 May 2018).

NASA (2017) 'Global climate change – vital signs of the planet' available at https://climate.nasa.gov/evidence/ (accessed: 21 June 2018).

Nature and Youth – Young Friends of the Earth Norway (2018) available at https://nu.no/english/ (accessed: 21 June 2018).

Nelson, Benjamin (1971) 'Discussion on industrialization and capitalism', in Stammer, Otter (ed.) *Max Weber and Sociology Today (Explorations in Interpretive Sociology)*, New York: Harper & Row.

New Scientist (2017) 'We are on track to pass 1.5°C warming in less than 10 years', available at www.newscientist.com/article/2130738-we-are-on-track-to-pass-1-5c-warming-in-less-than-10-years/ (accessed: 28 May 2018).

Our Children's Trust, Earth Guardians & The Global Initiative (2016) 'State obligations regarding Children's rights and climate change', *Submission to the UN Committee on the Rights of the Child*, available at https://meg-ward-wf6y.squarespace.com/s/OCT-et-al-CRC-Submission.pdf (accessed: 14 June 2017).

Oxfam Ireland (2018) 'Five critical things we learned from the latest IPCC report on climate change' available at www.oxfamireland.org/blog/5-things-climate-change (accessed: 20 June 2018).

Pettit, Philip (2005) 'The domination complaint', *Nomos*, 46: 87–117.

Pettit, Philip (2011) 'The instability of freedom as noninterference: The case of Isaiah Berlin', *Ethics*, 121(4): 693–716.

Rawls, John (2001) *Law of Peoples*, Harvard: Harvard University Press.

Ricoeur, Paul (2004) *Memory, History, Forgetting*, Chicago: University of Chicago Press.

Rosa, Hartmut (2015) *Social Acceleration: A New Theory of Modernity*, New York: Columbia University Press.

Samson, Kevin (2016) 'The privatization of water: Nestle denies that water is a fundamental human right', *Global Research* (29 August) available at https://counterinformation.wordpress.com/2014/11/06/the-privatization-of-water-nestle-denies-that-water-is-a-fundamental-human-right/ (accessed: 5 December 2018)

Sand, Peter. H. (2004) 'Sovereignty bounded: Public trusteeship for common pool resources', *Global Environmental Politics*, 4: 47–71.

Sassen, Saskia (2007) 'Response', *European Journal of Political Theory*, 6(4): 431–444.

Sheffield, P. E., Knowlton, K., Carr, J. L. & Kinney, P. L. (2011) 'Modeling of regional climate change effects on ground-level ozone and childhood asthma', *American Journal of Preventative Medicine*, 41(3): 251–257.

Short, Damien, Elliot, Jessica, Norder, Kadin, Lloyd-Davies, Edward & Morley, Joanna (2015) 'Extreme energy, 'fracking' and human rights: A new field for human rights impact assessments?' *International Journal of Human Rights*, 19(6): 697–736.

Sink or Swim (2018) 'South Florida sea level rise' available at http://miamisearise.com/
home-page/south-florida-sea-level-rise/ (accessed: 21 June 2018).

Skillington, Tracey (2015) 'Climate justice without freedom: Assessing legal and political
responses to climate change and forced migration', *European Journal of Social Theory*,
18(3): 288–307.

Skillington, Tracey (2017) *Climate Justice and Human Rights*, Basingstoke: Palgrave
Macmillan.

Snow, David & Benford, Robert (1988) 'Ideology, frame resonance and collective mobilization', in Klandermans, Bert, Kriesi, Hanspeter & Tarrow, Sidney (eds.) *From Structure to Action*, Greenwich, CT: JAI Press: 197–218.

Sorokin, Pitirim A. & Merton, Robert K. (1937) 'Social time: A methodological and func-
tional analysis', *American Journal of Sociology*, 42(5): 615–629.

UNICEF (2015) 'Unless we act now: The impact of climate change on children', November,
available at www.unicef.org/publications/index_86337.html (accessed: 30 April 2017).

UNICEF UK (2013) 'Climate change: Children's challenge' available at www.unicef.org.
uk/publications/climate-change-report-jon-snow-2013/ (accessed: 14 June 2017).

United Nations (2013) 'Intergenerational solidarity and the needs of future generations',
Report of the Secretary-General to the United Nations General Assembly, 15 (August)
available at https://sustainabledevelopment.un.org/content/documents/2006future.pdf
(accessed: 24 July 2017).

United Nations Convention on the Rights of the Child (1989) available at www.hrw.org/
news/2014/11/17/25th-anniversary-convention-rights-child (accessed: 14 June 2017).

United Nations Environment Programme (2010) 'UNEP emerging issues' available at
www.fao.org/fileadmin/user_upload/fsn/docs/HLPE/Environmental_Consequences_
of_Ocean_Acidification.pdf (accessed: 21 June 2018).

United Nations Environment Programme (2016) 'International resource panel, "Global
material flows and resource productivity"' available at http://wedocs.unep.org/handle/
20.500.11822/21557 (accessed: 2 June 2018).

United Nations Framework Convention on Climate Change (1992) available at https://
unfccc.int/resource/docs/convkp/conveng.pdf (accessed: 21 June 2018).

United Nations Paris Agreement (2015) available at https://unfccc.int/process-and-meet
ings/the-paris-agreement/the-paris-agreement (accessed: 21 June 2018).

United Nations Water (2018) 'Water scarcity' available at http://webcache.googleusercon
tent.com/search?q=cache:MNtcuD790KMJ:www.unwater.org/water-facts/scarcity/+&cd=
5&hl=en&ct=clnk&gl=ie (accessed: 1 June 2018).

United States Environmental Protection Agency (2017) 'Future of climate change' available at https://19january2017snapshot.epa.gov/climate-change-science/future-climate-
change (accessed: 28 May 2018).

Urry, John (2016) *What Is the Future?* Cambridge, MA: Polity Press.

Waldron, Jeremy (2002) *God, Locke and Equality: Christian Foundations of Locke's Political Thought*, Cambridge: Cambridge University Press.

Wallerstein, Immanuel (2012) 'Land, space and people: Constraints of the capitalist world
economy', *American Sociological Association*, XVIII(1): 6–14.

Weber, Max (1968) *Economy and Society: An Outline of Interpretive Sociology, Volume 3*,
eds. Guenther Roth & Claus Wittich, New York: Bedminster Press.

Weber, Max (1978) *Economy and Society: An Outline of an Interpretive Sociology, Volume
1*, eds. Guenther Roth & Claus Wittich, Berkeley: University of California Press.

Weizman, Eyal (2011) *The Least of All Possible Evils: Humanitarian Violence from Arendt
to Gaza*, London: Verso.

Wood, M. C. & Woodward, C. W. (2016) 'Atmospheric trust litigation and the constitutional right to a healthy climate system: Judicial recognition at last', *Washington Journal of Environmental Law & Policy*, 2(2): 634–684.

World Bank (2013) 'Thirsty energy: Securing energy in a water constrained world' available at www.worldbank.org/en/topic/sustainabledevelopment/brief/water-energy-nexus (accessed: 25 May 2018).

World Bank (2014) 'Grey rock – transforming natural gas' available at www.greyrock.com/wp-content/uploads/2014/10/Greyrock_Mini_GTL_WorldBank_Report.pdf (accessed: 19 August 2017).

World Bank (2016) 'Sustainable development and the demand for energy' available at https://blogs.worldbank.org/opendata/sustainable-development-and-demand-energy (accessed: 16 November 2017).

World Energy Council (2013) 'World resources 2013 survey' available at www.worldenergy.org/wp-content/uploads/2013/09/Complete_WER_2013_Survey.pdf (accessed: 25 June 2017).

World Future Council (2013) 'Crimes against future generations' available at www.worldfuturecouncil.org/crimes-against-future-generations/ (accessed: 26 June 2017).

World Health Organization (WHO) (2017) 'The cost of a polluted environment: 1.7 million child deaths a year, says WHO' (6 March) available at www.who.int/mediacentre/news/releases/2017/pollution-child-death/en/ (accessed: 17 May 2017).

World Meteorological Organization (2017) 'Greenhouse gas concentrations surge to new record' available at https://public.wmo.int/en/media/.../greenhouse-gas-concentrations-surge-new-record (accessed: 21 June 2018).

World Nuclear Association (2016) 'World nuclear performance report 2016' available at http://world-nuclear.org/getmedia/b9d08b97-53f9-4450-92ff-945ced6d5471/world-nuclear-performance-report-2016.pdf.aspx (accessed: 26 June 2017).

World Wildlife Fund (2018) 'Water scarcity' available at www.worldwildlife.org/threats/water-scarcity (accessed: 1 June 2018).

Zhiwei, Xu, Sheffield, Perry, Su, Hong, Wang, Xiaoyu, Bi, Yan &Tong, Shilu (2014) 'The impact of heat waves on Children's health: A systematic review', *International Journal of Biometeorology*, 58(2): 239–247.

Chapter 2

Changing the evaluative discourse on climate change

The campaign for future justice

Introduction

Having laid out some of the ways in which relations between generations are marked by deep inequalities in Chapter 1, this chapter explores how this situation is subject to critical interpretation by future justice campaigners. It notes a high degree of convergence in the thinking of future and youth justice coalitions on the way principles of justice and right (e.g., to live resilient, peaceful and sustainable lives) are being compromised by the expanding energy needs of global capitalism and the desire to challenge the illegitimate nature of power yielded by global elites today over the ecological fate of billions. The conditions necessary to preserve a just basic resource structure, including net investments in and cumulative gains from material, cultural, educational and technological resources but, most importantly of all, unpolluted natural resource reserves are not being protected for the future. Campaign actors highlight the dangers of this situation by bringing together an interpretation of crisis as it arises and is experienced in everyday life with a transcendent constructive moment, when cumulative problems are framed as human rights violations requiring immediate democratic intervention (Klandermans, 1984; Snow & Benford, 1988; Strydom, 2011). In that, these actors may be said to offer a valuable critical reconstruction of climate change issues as matters of deep justice concern. They draw publics' attention to the manner in which obligations to share nonrenewable resources with those that come later are losing their institutional potency and consider how this situation might be reversed. Campaigners draw on a range of 'tools of interpretation' when addressing these issues, including existing repertoires of protest (e.g., rights claims) and frames of meaning whose 'cultural conduciveness' (Gamson, 1988, 1992) is amplified by growing direct evidence of ecological deterioration (e.g., the loss of ecosystems and the accumulation of toxins in fish stocks, crops, plants and animal food chains; see, for example, UNEP Frontiers 2016 Report).

The following analysis considers the importance of these actors' interpretive work to a public steering of these issues towards more democratic processes of crisis response. In that, it acknowledges the 'signifying' role of these actors as 'meaning constructing agents' (Snow & Benford, 1988, 2000) in climate change discourse today. It notes the significance of these actors' assessments of the way ongoing deteriorations in global ecological conditions make the determination of equitable treatment across generations increasingly difficult. Even allowing for the challenges that declining environmental conditions create for the present, duties to invest in the future, they argue, still hold (see also World Resources Institute, 2014). If future generations are to have sufficient opportunities to adapt to what are certain to be harsher living environments in the years ahead (Reiman, 2007: 80; Tremmel, 2014: 199), then resources must be conserved for their benefit. A stronger institutional commitment to intergenerational equity may not produce perfect equality across time for several reasons, some even possibly at this stage beyond our control (e.g., existing levels of ozone pollution). However, the imperative to act ought to be self-evident (Barry, 1999: 107; Hösle, 2003), yet, it seems, it is not. Obligations to protect the ecological inheritance of newer generations are being ignored (World Future Council, 2013). Campaign actors identify, reconstruct and make explicit the reasons why this problematic situation arises, noting, in particular, how primary polluting agents today violate relations of trust, mutual respect and long established justice principles (e.g., the principle of 'just savings', see Rawls, 1999: 159; McKinnon, 2012) in ways that threaten all our future.

Framing climate change at close range

Those who draw attention to the importance of these issues include the Foundation for the Rights of Future Generations, the World Future Council, Earth Guardians, Foundation for Democracy and Sustainable Development, Our Children's Trust and Future Justice. The following analysis explores how these actors interpret what a greater intergenerational equity actually requires, as well as the forces currently blocking society's capacities to realize commitments to a sustainable future. Broadly speaking, these actors bring together a diagnostic critique of what is fundamentally wrong with current human relations under a global capitalist regime (a problematic present) with a transcendent constructive moment of democratic reform (future alterity). Certain common patterns of interpretation may be detected across these actors' framing efforts. For instance, they all challenge dominant discursive formulations of environmentally destructive practices as 'unavoidable' side effects of humanity's progression from less to more advanced living conditions (e.g., Future Justice, 2016). Second, all amplify reasons why the fossil fuel economy is a major contributor to climate change yet continues to be generously supported by government and international financing agencies. All highlight how the wealth and pollution generated by this economy are unevenly distributed and predicated on major divisions between peoples, explored through themes of inequality and exploitation (e.g., see the Fossil Fuel Divestment Student Network, 2017; Also, Friends of the Earth Europe, 2009). When experienced as scorching heat waves, severe water shortages, flash flooding, storm surges and drought, climate change compels a more critical look at its underlining causes, as well as more conventional formulations of nature's destruction as a human right (Future Justice, 2016). For youth, especially, this critical process of interpretation occurs from the vantage point of those whose future has been stolen and whose right to participate in decision making on various policy issues is denied. The cogency of traditional justice beliefs is, therefore, critically reassessed from the standpoint of the democratically dispossessed.

In terms of diagnostic framing (Büchs et al., 2015), blame is placed firmly at the feet of the fossil fuel industries and those states that fail to control their activities. Both are accused of 'foreclosing' on humanity's future (see, for example, Occupy Wall Street, 2011a) and 'tearing apart' the life-giving ecosystems 'upon which we depend' (Movement Generation, 2011) 'in the service of higher profits' (Global Justice Ecology Project, 2016). One image produced by the Foundation for the Rights of Future Generations (2003: 13) gives expression to this evaluative discourse. Capitalism is personified in the figure of the oil tycoon who burns an 'inter-generational contract' to preserve resources for generations to come, while the ordinary people of this world struggle to prevent planetary destruction.

Visually, the theme of foreclosure is also explored through the figure of the corporate executive or banker (see also Figure 2.1) writing the rules of an unfair economic order that threatens communities everywhere.

Rhetorically, the theme of foreclosure is discussed in relation to specific examples. The World Future Council (2016), for instance, focuses on the case of Bolivia where, oil and gas companies have been granted exploration rights to the lands of 11 of the state's 22 protected areas, including 75 percent of the Madidi National Park in Northwest Bolivia (home to 11 percent of the world's species of birds). Legal arrangements granting unprecedented freedoms to corporations to 'radically modify protected areas' of the world are condemned, as are rules advancing the interests of major corporations, such as new bilateral investment rules, free trade agreements, land leasing arrangements and a system of dispute settlement that favors the interests of big business (Hill, 2015). Particular concern is expressed at the way larger corporations continue to strengthen their legal capacity to gain access to remote resource reserves that in their natural state (gas and oil reserves, ocean minerals, etc.) belong to all in common. Attention is focused on the current absence of 'hard' legal instruments capable of compelling corporations to address damages caused by their manufacturing activities and to respect human rights law. In response to these developments, the World Future Council (2016) calls for the introduction of a tighter legislative control of the operations of transnational corporations and greater institutional efforts to address the inequalities that prevail between the economic and political bargaining power of major polluters and ordinary communities everywhere. Similarly, the Foundation for Democracy and Sustainable Development (FDSD, 2013) draws attention to the present absence of a sufficiently just order of regulatory control:

We cherish democracy: the rule of the people, by the people, and for the people. But democracy is undermined by decision-making that is democratic in name only. It is threatened by conflict, apathy, inequality, manipulation, and corruption. It is failing to deliver sustainability.

(FDSD, 2013)

Figure 2.1 Occupy Your Future

Source: Image designed by Hello Cool World, courtesy of TheCorporation.com (2018).

We call special attention to the fact that human rights violations by transnational corporations and other business enterprises – often in complicity with the local governments – not only deprive people of their human rights today but also deprive future generations of their rights.

(World Future Council, 2016)

In recent years, a number of moves have been made to draw attention to these problems. In 2014, the UN Human Rights Council, for instance, adopted resolution A/HRC/RES/26/9, insisting that international human rights law be better used to regulate the activities of transnational corporations and ensure their strict compliance with legal norms. This was followed by the establishment of a Working Group on Business and Human Rights to encourage and promote 'capacity-building' with regard to the implementation of the UN's Guiding Principles on Business and Human Rights (see Office of the High Commissioner for Human Rights, 2018). UN agencies play a key role in opening up potentially critical discourse spaces and introducing 'topic markers' (van Dijk, 1977) in human rights violations to which nongovernmental organizations (NGOs) can respond, furthering public communication on such issues. Greenpeace International (2017), for example, questioning why governments do not, in line with the UN's recommendations, challenge 'big polluters' or subject their operations to more stringent regulatory controls (World Future Council, 2017)? They draw attention to the, at most, piecemeal efforts of states to date to offset the effects of escalating environmental problems and the dangers of such in a world that is increasingly divided and overheated (see, also, FDSD, 2017). Mindful of the fact that most legal obligations designed to protect the welfare of future generations are being seriously challenged at present (by rising atmospheric pollution and government inaction), these actors call on states to upgrade their efforts to control carbon emissions levels and radically reduce rates of natural resource depletion.

The Foundation for Democracy and Sustainable Development (FDSD, 2011) as well as Climate Justice Now (2016) remind states of the need to respond to the current ecological crisis 'as if the future matters, while there is time' (Elkington, 2011). Deliberate obstructions to the regulation of carbon production on the part of industry, they add, must be controlled by states if safe living conditions are to be secured for the future. With similar concerns in mind, the Human Rights Council adopted Resolution A/HRC/35/13 in 2017, noting the disproportionate impact of climate change (increased frequency of natural disasters, food and water shortages and the transmission of disease) on the rights of the child:

Human rights commitments contained in the Convention on the Rights of the Child and the Paris Agreement make clear that States have affirmative obligations to take action to protect the rights and best interests of the child from the actual and foreseeable adverse effects of climate change. Failure to take adequate steps to prevent children from suffering foreseeable climate-related human rights harms breaches these obligations.

(OHCHR Summary of the Analytical Study of the Relationship between Climate Change and the Full and Effective Enjoyment of the Rights of the Child, 2017)

Framing a lack of affirmative action on the part of states as equivalent to a breach of legal obligations is one that resonates with but also offers further affirmation to the arguments presented by future climate and youth justice coalitions. Together with references to established legislation affirming the 'inherent dignity' of all members of humanity (including the Aarhus Convention on Access to Information, Public Participation in Decision-Making and Access to Justice in Environmental Matters [Article 1, convention *entered into force* 30 October 2001], the Declaration of Bizkaia on the Right to the Environment, 1999–UNESCO, the International Covenant on Economic, Social and Cultural Rights [1966], the International Covenant on Civil and Political Rights [1966], and the Convention on the Rights of the Child, 1989), this discourse raises the idea of intergenerational justice to the level of a legal and political event, that is, one through which ideative and symbolic constructs of justice are rendered more societally relevant.

For youth campaigners, however, the real source of change is not governments as such but people. It is people who have the potential to remake relations of inequality (Earth Guardians, 2016). Attention must, therefore, shift to 'those of us who share the goals of liberation and a workable future for our children. There can be no hierarchies of oppression' (Our Generation, Our Choice, 2015). 'We are the ones we've been waiting for to lead projects, campaigns, and collective actions all over the planet' (see Climate Generation, 2017). Emphasis is repeatedly placed on the capacity of communities to take action and consciously shape the direction of their future (what the sociologist Alain Touraine (1977) refers to as 'historicity'): 'People power is the only thing that has ever created change and we are unstoppable when we stand together' (Yong Jung Cho et al., 2015). It is 'for the world community to decide', these actors argue, whether this phase in the history of the Anthropocene will tolerate deepening inequalities, relations of domination, suffering and the erosion of the rule of law or, alternatively, will renew institutions of democracy and commitments to human rights (see, for example, 'Our Generation Our Choice' – Mobilization for Justice on Race, Climate Change and Immigration, 2015).

Our civilization has never faced such existential risks as those associated with global warming, biodiversity erosion and resource depletion. Our societies have never had such an opportunity to advance prosperity and eradicate poverty. We have the choice to either finally embark on the journey towards sustainability or to stick to our current destructive 'business-as-usual' pathway.

(Earth Statement, 2015)

Ultimately, the realization is that a more equitable and ecologically sustainable world cannot be some future project of action but must be located in the energies, the experiences and critical reflexive capacities of the present. The present moment still offers a window of opportunity to decarbonize the world and initiate a movement towards greater intergenerational equity and ecological stewardship (see, for example, Foundation for Democracy and Sustainable Development, 2009). However, the understanding is also that intergenerational justice will gain real societal momentum only if it becomes a set of 'ideas in action' (FDSD, 2009), that is, one that has the genuine backing of publics, political decision making and administrative support systems. To gain the necessary degree of support, intergenerational justice must first be seen by all to be serving humanity's 'intrinsic' interests, that is, concerns that outweigh all others in terms of long-term survival (Westall, 2015).

Social representations of intergenerational wrongdoing

Nowhere are these issues explored more extensively than through social media (Byrne, 2015). Social media provides these actors with an indispensable tool with which to devise new 'infra-national maps' (Beck, 2003: 467) of cosmopolitan belonging, where youth, in particular, evaluate the micro details of their generational location from the shared vantage point of the excluded. Communication platforms, such as Facebook and Twitter, allow for a more regular engagement with themes of justice, inequality and rights (Karpf, 2010). These developments have been noted by several authors (e.g., Bennet, 2003; Bennet and Segerberg, 2011; Valenzuela, 2013; Scherman, Arriagada and Valenzuela, 2014), as has Millennials' tendency to situate protest politics at the intersection of several pressing concerns at once (e.g., climate, immigration, economic crisis and racial injustice, for instance) (Barberá et al., 2015). In general, however, social science research has not yet adequately explored shifting intersections in the issue agenda of contemporary protest actors (Franco & Zimbardo, 2006) or even how a cosmopolitan imagination inspires these activists' contemporary global agenda. 'Our movements are connected globally and so is our purpose' (350.org., 2015). As Yong Jung Cho, Waleed Shahid, Devontae Torriente and Sara Blazevic (Our Generation, Our Choice Action Team, 2015) explain, across the world 'young people don't live single issue lives so why would we pursue single-issue political change'? The ensuing analysis examines some of the ways in which themes of intergenerational justice are explored through this genre of political representation, how ecological, economic, and social crises are examined through written and visual narratives of protest. In the case of visual narratives, the viewer is drawn in as a witness to re-presented scenes of ecological destruction, economic exploitation and political injustice. What features regularly throughout is the image of the child, used symbolically as a bridging device to connect the destructive tendencies of the present with the diminished life chances of future generations. In the process, questions of justice are rendered more concrete, and the 'distant suffering' of future peoples is brought closer to home.

The image of the child is employed as an interpretive tool (Donati, 1992: 141) to make sense of a complex reality (climate change) in terms of the everyday. 'Amorphous, ill defined problematic situations' (Rein, 2012: 269), such as environmental pollution, storm surges, flooding, etc., increasingly, invade the 'normal'. As indicators of environmental destruction, these evoke the perceiver's existing knowledge of the world and allow her/him to construct a meaningful understanding of ecological changes as they occur. Frames or 'schema of interpretation' (Goffman, 1974) thereby help to 'conventionalize' climate change as a real and unfolding social experience. Future justice campaigners attempt to direct this process of cognitive framing towards the primary victims of ongoing environmental degradation – youth.

Emphasis is placed on the vulnerabilities of the child to growing ecological adversities and exploitation. The child has 'done least to contribute to the release of greenhouse gases', as Intergenerational Foundation (2016) points out, yet will suffer most from its consequences. Emphasis is placed on obligations to protect the ecological inheritance of children, especially from those who attempt to privatize its essential components and individualize responsibility for the endangerment of our future. The hope is to deepen publics' moral identification with and emotional connection to future peoples via the figure of the child. Typically, the innocent child is juxtaposed against a background of escalating ecological destruction or symbols of economic exploitation. The act of exploitation is rendered 'knowable' with the aid of series of visual cues (e.g., the businessman with a briefcase in the case of *Occupy Your Future*, see Figure 2.1). Rhetorically, also, emphasis is placed on our responsibility as citizens, as guardians, and as protectors of our Children's future to address growing threats of ecological disaster:

> For my generation and all those to follow, nothing can be prioritized above the climate crisis. Your actions will shape the world you will leave to your grandchildren and all future generations and after you are gone, we will still be here. What legacy will you leave us?
>
> (Xiuhtezcatl Martinez, Earth Guardians, June 2016)

The transnational appeal of this communication style helps to reinforce a cosmopolitan ethic of solidarity amongst all communities, especially youth. Participants in this communication are encouraged to think more deeply about the social, economic and political processes in which their own partial experiences of climate injustice are embedded. More 'enlarged' modes of understanding (Young, 1997) climate change as the product of relations of inequality and denied opportunities to realize sustainable worlds are actively encouraged. In particular, climate change comes to be construed as an injustice against future generations (see World Future Council, 2016).

In the visual imagery of protest, such as that produced by the Foundation for the Rights of Future Generations (2003: 10), the dual role of generations as beneficiaries of the conservations efforts of past generations and as trustees of future generations' resource needs is explored, albeit ironically. The understanding is that relations of trust between generations are abused by those who continue to misspend future peoples' resource capital. Similarly, in a visual produced by Hello Cool World, courtesy of The Corporation.com for Occupy (2017), a child is depicted as claiming entitlement to a safe and democratic future (holding a banner with the words 'occupy your future'), but her interests and welfare are compromised by what is figuratively represented here as a 'dark male presence' behind her (holding a briefcase). Focusing on the child whose future overlaps with generations not yet born increases the appeal of an intergenerational perspective on justice and, simultaneously undermines efforts to distance present wrongdoing from its long-term effects. Within a temporally extended commons, solidarity with those located at different points across 'a chain of intergenerational bonds' (Gillot Assayag, 2016) is actively encouraged 'An ethos of generosity' (Brigstocke, 2016: 161) is promoted rather than one of austerity. The understanding is that the interests and entitlements of all generations must be protected.

One protest message used by Occupy during Wall Street protests in September 2011 was the following: 'Dear Future Generations, Please accept our apologies. We were rolling drunk on petroleum'. By evoking a theme of intergenerational wrongdoing, Occupy drew attention to ecological debts owed to future generations, that is, debts accrued through a knowing imposition of environmental harms that continue to grow in parallel with rising carbon emissions. To relieve the burden of such debts, remaining fossil fuel reserves must be preserved and the commons protected. Underlining the notion of present polluters being indebted to future generations is a theme of irresponsible behavior. The latter is reinforced through a careful choice of descriptive wording; for instance, 'rolling drunk on petroleum' (Occupy Wall Street, 2011a) or 'bingeing on fossil fuels' (Friends of the Earth, 2011), behaviors that are said to require immediate corrective action. These actors are not content with simply clarifying what obligations are owed to future generations. The intention also is to bring about real social, ecological, legal, political and economic change. The challenge, therefore, is to act to realize greater equity in the present (World Future Council, 2016). Youth-led climate justice movements, in particular, advocate this view. Earth Guardians (2016) envisage their role as empowering and elevating 'the voices of rising generations of climate solutionaries'. Xiuhtezcatl Martinez (2016), executive youth director of the group Earth Guardians, describes himself as an indigenous environmentalist, a student, eco hip-hop artist and concerned citizen 'on the frontlines of a youth-led climate movement since he was six years old', traveling across continents to Africa, India, Australia, Europe, New Zealand, bringing a message of empowerment:

We have the solutions at our fingertips, we have the innovation, the creativity, and the passion that we need to bring forth a healthy, just, sustainable world for every generation to come. Regardless of what happens, the people are ready to carry forth the momentum, the energy, and the movement necessary to change the world after COP 21 [the 2015 United Nations Climate Change Conference].

(Democracy Now, April 14, 2016)

Martinez's words capture the essence of a new determination amongst youth activists today to reclaim their ecological destiny from those who 'do not seek their consent' to extract their wealth and the Earth's (Occupy Wall Street, 2011b) and realize their identity as agents of democratic justice (Indian Country, 2013). In 2012, Martinez spoke at the Rio+20 United Nations Summit in Rio de Janeiro to panel representatives about the concerns and aspirations of contemporary youth. In 2013, he received a United States Community Service Award and was chosen to serve on President Barack Obama's youth council. It is younger activists especially, such as Martinez, who highlight the importance of the future to the political present and express deep concern about their exclusion from the decision-making process.

Asserting rights to democratic participation

In their 2016 submissions to the UN Committee on the Rights of the Child, a number of campaign groups, including Our Children's Trust, Earth Guardians and The Global Initiative drew attention to states 'repeated and direct' violation of international legal obligations, not only to consider but also include children (representatives) in decision making on climate change matters. The Paris Agreement notes the importance of extending unique consideration to children, as a special group, particularly vulnerable to the threats posed by deteriorations in climate conditions. Article 2 of the CRC obliges states to enforce the requirement to 'respect and ensure the [CRC enumerated] rights . . . without discrimination of any kind, irrespective of the child's or his or her parent's or legal guardian's race, color, sex, language, religion, political or other opinion, national, ethnic or social origin, property, disability, birth or other status'. Under Article 3(1) of the CRC, states are obliged to put the 'best interests' of children at the top of their issue agenda. Also, Article 12(2) obliges states to ensure that children are 'provided with the opportunity to be heard in any judicial and administrative proceedings affecting the child, either directly or through a representative or an appropriate body'. In establishing Young and Future Generations Day on December 3, 2015, and encouraging Children's concerns to be brought more to the fore, the COP 21 not only acknowledged the contribution of children to climate change debate but also the legitimacy of their right to be party to the same (see UNFCCC, Young and Future Generations Day, 2015). The importance of making the voice of youth heard is brought home by the fact that 85 percent of the world's youth population currently live in developing regions of the world where in many instances average surface temperatures are expected to increase more than 2°C in the decades ahead. The dire ecological effects of such temperature rises will disproportionately affect children and other vulnerable sections of the population (Our Children's Trust, 2016). States are legally obliged to ensure opportunities to adapt to these ecological conditions will be available to all (UNFCCC, 1992; Kyoto Protocol, 1997).

The most explicit form of discrimination exercised against children, according to Our Children's Trust and Earth Guardians, is that performed by states in their systematic failure to control levels of GHG emissions. States continue to make policy decisions that benefit the short-term interests of a few (e.g., the ongoing subsidization of fossil fuel industries), whilst ignoring the astronomical costs of these decisions for the majority. Costs are dispersed amongst a global population in the form of growing risks to health, environment, community, and future opportunity (Currie, 2016; UNICEF UK, 2013). For such reasons, campaigners insist that youth 'ought to be central to debates [about climate change] because they – as well as future generations – have a much larger stake in the outcome' than older generations (Currie, 2016: 4). Through political campaigns, legal actions and participation in hearings, youth, together with their supporters are pressurizing states to recognize and protect their rights and challenge the legitimacy of policies driving further environmental destruction (Our Children's Trust, 2017). Youth, in collaboration with Our Children's Trust, are taking legal actions at the municipal, state, and federal level, directly invoking legal obligations to restore a stable climate system and protect all peoples' fundamental rights (see Chapter 4 for a more detailed account of these developments). Relevant legal instruments include the Convention on the Rights of the Child, which references four varieties of right: survival rights, development rights, protection rights and participation rights. Article 12(1) of the Convention requires states to allow children to express their 'views freely in all matters affecting' them, while Article 12(2) requires that children be provided with the 'opportunity to be heard' during administrative procedures. Article 17 obliges states to provide children with information on their legal rights and a context to communicate ideas on mitigation, prevention and response planning for current and future ecological threats. In other words, youth should not have to take states' representatives to court to have a say in a debate in which they have a right to participate.

To counter the injustice of their exclusion from decision making on climate change issues, increasing numbers of youth call for a lowering of the voting age to 16 years. Youth point to the fact that they have a fundamental interest in how crisis issues are addressed and demand a right to participation in procedures of democratic decision making. The fact that youth do not speak with one voice or represent just one set of interests increases the necessity of their inclusion in decision-making procedures. Not only are there social, economic and political inequalities between them, but such differences will also become increasingly evident in the decades ahead as climate conditions continue to decline. Differences among youth are already evident in terms of the exposure to climate change harms and the capacity to adapt to ecological adversities (depending on regional location, economic status, etc.). Ensuring a wide participation of such varying interests is, therefore, essential (Smith, 2015).

> The most severely under-represented interests in our political system are young people under the age of 18 because they cannot vote. . . . A new Bill

of Rights must be designed to ensure that future generations have reasonable prospects for freedom, dignity, prosperity, financial stability and survival.

(Earth Manifesto, 2015)

There are 74 million of us under 18 [in the United States] who cannot be a part of our democratic process but we will be the beneficiaries or the victims of whatever you do or don't do in relation to our current climate emergency.

(Earth Guardians, 2016)

Transforming the future through the present

In the process of trying to realize more sustainable and democratic futures, something else is produced (see Bloch, 1995) – a socially engaged and politically aware youth. A sustainable commons is produced by those who are made in the process ('*processus cum figures, figurae in processu*' Bloch, 1995). The rise of the contemporary youth activists (amongst other future justice campaigners) forces sociology to revise its portrayal of the champions of democratic reform. The tendency has been to ignore the contribution of youth (Anderson, 2016), yet the degree to which they mobilize legally and politically today in response to major climate justice issues cannot be overlooked. Rising Youth for a Sustainable Earth (RYSE), 350.org, UK Youth Climate Coalition and other 'kid warriors' play an increasingly prominent role internationally, highlighting the shared nature of ecological challenges and the fact that climate solutions, as a subject of a collective willing, cannot be limited de facto to members of one particular community, in generational or territorial terms. As they do so, daily encounters with environmental problems and those commonly associated with climate change are subject to a more critical process of interpretation, one mediated by cultural understandings of rights, duties and responsibilities. The preceding analysis considered how campaigners contribute to this process of reinterpretation by focusing critical attention on key factors, such as the issue of responsibility. Second, by facilitating new mechanisms of articulation (e.g., tying together various accentuated elements of the 'why', 'how' and 'who' story of climate change), these actors draw public attention to what was previously unexplored. Third, these actors perform 'a transformative function' (Snow, 2013), encouraging society to think more creatively (with the aid of modernity's traditions of democracy, peace and human rights) about how we can move forward in a more sustainable manner.

Youth offer a compelling critique of the dominant evaluations of climate change (e.g., that espoused by the state), its consequences, as well as its causes. However, they offer a prefiguration of the current democratic moment, an account of how the future can be democratically reformed through the present. Their aim is to 'change the landscape of what's politically possible' (Yong Jung Cho et al., 2015) and to reverse trends towards the foreclosure of economic, social, political and ecological worlds. Youth call on publics to 'occupy the future!' 'Now is the time to act. . . . It is within reach. From pragmatic beginnings could emerge a visionary change of direction for the world' (UN, 2005: 6). The cosmopolitan dimensions of these actors' demands are without borders or membership requirements. The global other is internalized in a solidaristic manner, as campaigners search for the transformative potential of the current democratic moment: 'we are all human beings first and privileged with responsibilities to each other, to future generations and to the planet' (Avaaz, 2011).

The 'impossibility' of alternative worlds is rigorously contested and new cooperatively produced options explored instead. The 'slow movement', for instance, attempts to change how we live in and relate to the world around us. It promotes obduracy (persistence through time) or occupation of spaces for long periods, insisting upon peoples' right to be present (and their voice heard) in the shaping of social, ecological and political futures. The objective is to open up new interpretive, social and political spaces where time can be socialized and shared as a reflective, critical endeavor rather than as a by-product of capitalist efficiency and work performativity. Similarly, Earth Guardians (2016) promote ways of living that encourage people to 'walk lightly and respect' the Earth's life support systems by consuming less and at a slower pace. Other movements, such as the Transition Town Movement, encourage society to 'transition' to a model of smaller, more self-sufficient community living. Here the emphasis is on local energy sources (wind, solar, hydro energy) that are collectively managed and owned by the people, substituting global with localized goods and services economies, fossil fuel energy needs with a series of alternative, zero-carbon energy, transport and consumption options.

> Our central survival task for the decades ahead, as individuals and as a species, must be to make a transition away from the use of fossil fuels – and to do this as peacefully, equitably and intelligently as possible.
>
> (Transition United States, 2017)

The Transition Movement affirms the abilities of local communities to 'step into a leadership position' on climate change and begin to work towards mitigating the interrelated effects of peak oil, climate change and economic crisis before it is too late. Collectively, what these actors offer is a new explanatory language with which to understand the problems created by existing institutional arrangements, as well as a blueprint for the initiation of less ecologically challenged and intergenerationally equitable futures. In the process, the idea of extending justice across generations is rendered more achievable and reasonable. However, it remains heavily dependent on securing a sufficient level of public support for transgenerationally grounded justice imperatives where primary emphasis is on a proportionally balanced pursuit of economical, ecological and social targets. This may mean taking actions that prove somewhat disadvantageous to the present but that are beneficial to humanity in the long term. Securing public support for such measures will not be an easy task, however, especially as the attitudes and practices of many citizens today are heavily implicated in the very dysfunctionalities that short-term reasoning gives rise to (e.g., convenience lifestyle choices).

According to Ulrich Beck (2015), the more likely scenario is one where a long-term perspectivism on justice emerges gradually on the basis of more frequent encounters with ecological disaster (e.g., more regular extreme weather events, periods of water crises, etc.). Direct experiences of ecological catastrophe and essential resource deprivation will eventually induce a shift in cognitive perspective and increase sensitivity to the need for more preventative action. Beck describes how the 'anthropological shock' (Beck, 2015) of violent storms, flash flooding, severe drought and so forth will increase awareness of our vulnerability to the natural environment, an awareness that eventually will translate as a desire to address primary sources of harm. Certainly, there is growing evidence of communities adopting a more proactive stance in relation to efforts to control climate change. However, there is also evidence to suggest that many are still unconvinced of the benefits of these practices, especially reducing current rates of depletion of essential nonrenewable resources. For global corporate actors and wealthier states, the priority has become taking all necessary steps in preparation for likely future disaster. 'Anthropological shock' in this instance does not translate as a pursuit of more sustainable practices or a desire for greater cooperation with others. Rather, it leads to what many describe as a 'global resource grab' (Amanor, 2012), with these actors stepping up efforts to source alternative supplies of cheap land, coal, oil, crops and water reserves abroad in the interests of preserving high profit margins.

Alarmed by governments' lack of urgency with regard to this state of affairs or commitment to a genuine program of sustainable development, campaigners consider how a more effective response might be developed. Earth Guardians (2017), for instance, call for the implementation of a new 'intergenerational contract' where present generations are legally prohibited from generating excessive carbon, overpolluting rivers and oceans or depleting soil nutrients excessively (see also Göpel, 2011). The following chapter considers the moral, legal and political components of this and similar approaches, as well as those of arguments against them. It offers a detailed account of the main points presented in opposition to assigning a legal status, including rights, to peoples not yet born, taking into consideration the practical, as well as moral objections raised by a variety of actors. It then lays out some of the arguments presented in defense of such moves (e.g., defining environmental security as a public good held across generations or each generation as a temporary possessor of agency rather than the sole beneficiary of the world's resource wealth). Steered by a greater appreciation for knowledge of harm, traditions of rights, duties and responsibilities are called upon increasingly to deal with a growing range of inequalities between peoples across regions, states, generations, and so on. As they do, the question of which social cleavages are fueling inequalities is opened up for wider debate. Sociologically, these developments are important on account of the contribution they make to a deeper societal engagement with issues of inequality and the human rights dimensions of climate change. The analysis that follows offers a detailed exploration of this process of engagement, as it unfolds across a range of social actors today.

References

350.org (2015) 'Why (we as) climate activists stand with refugees' available at https://350. org/why-we-as-climate-activists-stand-with-refugees/ (accessed: 23 June 2018).

Aarhus Convention on Access to Information, Public Participation in Decision Making and Access to Justice in Environmental Matters, available at ec.europa.eu/environment/ aarhus/index.htm (accessed: 23 June 2018).

Amanor, Kojo Sebastian (2012) 'Global resource grabs, agribusiness concentration and the smallholder: Two West African case studies', *The Journal of Peasant Studies*, 39(3–4): 731–749.

Anderson, Priscilla (2016) *The Politics of Childhood Real and Imagined, Volume 2*, Oxon: Routledge.

Avaaz (2011) available at www.avaaz.org/page/en/about/ (accessed: 16 October 2017).

Barberá, Pablo, Wang, Ning, Bonneau, Richard, Jost, John T., Nagler, Jonathan, Tucker, Joshua, Gonzalez-Bailon, Sandra. (2015) 'The critical periphery in the growth of social protests', *PLoS ONE*, 10(11): e0143611. doi: 10.1371/journal.pone.0143611

Barry, Brian (1999) 'Sustainability and intergenerational justice', in Dobson, Andrew (ed.) *Fairness and Futurity*, Oxford: Oxford University Press: 93–117.

Beck, Ulrich (2003) 'Toward a new critical theory with a cosmopolitan intent', *Constellations*, 10(4): 454–468.

Beck, Ulrich (2015) 'Emancipatory catastrophism: What does it mean to climate change and risk society?' *Current Sociology*, 63(1): 75–88.

Bennett, W. Lance (2003) 'New media power: The internet and global activism', in Couldry, Nick & Curran, James (eds.) *Contesting Media Power*, New York: Rowman & Littlefield: 17–38.

Bennet, W. Lance & Segerberg, A. (2011) 'Digital media and the personalization of collective action: Social technology and the organization of protests against the global economic crisis', *Information, Communication & Society*, 14: 770–799.

Bloch, Ernest (1995) *The Principle of Hope: Volume One*, Cambridge, MA: MIT Press.

Brigstocke, Julian (2016) 'Occupy the future', in Kirwan, Samuel, Dawney, Leila & Brigstocke, Julian (eds.) *Space, Power and the Commons: The Struggle for Alternative Futures*, London: Routledge: 150–165.

Büchs, Milena, Saunders, Clara, Wallbridge, Rebecca, Smith, Graham & Bardsley, Nicholas (2015) 'Identifying and explaining framing strategies of low carbon lifestyle movement organizations', *Global Environmental Change*, 35: 307–315.

Byrne, Colin (2015) 'Getting engaged? The relationship between traditional, new media, and the electorate during the 2015 UK general election', *Reuters Institute for the Study of Journalism* (November) available at http://reutersinstitute.politics.ox.ac.uk/publi cation/getting-engaged-relationshipbetween%20Traditional,%20Media%20and%20 the%20Electorate.pdf (accessed: 7 March 2017).

Climate Generation (2017) 'We are the ones we've been waiting for' available at www. climategen.org/blog/ones-weve-waiting/ (accessed: 22 June 2018).

Climate Justice Now (2016) 'Our land is worth more than carbon', available at www.tni. org/en/article/our-land-is-worth-more-than-carbon (accessed: 23 June 2018).

Currie, Janet (2016) 'Children and climate change', *The Future of Children*, 26(1): 3–10.

Declaration of Bizkaia on the Right to the Environment (1999) 'UNESCO' available at http://unesdoc.unesco.org/images/0011/001173/117321E.pdf (accessed: 11 August 2017).

Democracy Now (2016) 'Landmark climate lawsuit: Meet the youth activists suing the U.S Government & Fossil Fuel Industry' available at www.democracynow.org/2016/4/14/landmark_climate_lawsuit_meet_the_youth (accessed: 5 December 2018).

Donati, Paolo (1992) 'Political discourse analysis', in Diani, Mario & Eyerman, Ron (eds.) *Studying Collective Action*. London: Sage: 136–167.

Earth Guardians (2016) available at https://www.givetwig.org/blog/2016/3/8/xxx. (accessed: 5 December 2018).

Earth Guardians (2017) 'Pledge to be a climate leader' available at www.earthguardians.org/youthpledge/ (accessed: 11 August 2017).

Earth Manifesto (2015) 'A bill of rights for future generations' available at www.earthmanifesto.com/1%20-%20A%20Bill%20of%20Rights%20for%20Future%20Generations.htm (accessed: 23 June 2018).

Earth Statement – Global Challenges Foundation (2015) available at https://globalchallenges.org/our-work/earth-statement-2015/statement (accessed: 23 June 2018).

Elkington, John (2011) 'We need a ministry for future for future generations for our Children's sake', *The Guardian*, 23 November.

Fossil Fuel Divestment Student Network (2017) 'Our public lands and waters are under threat' available at www.patagonia.com/actionworks/campaigns/public-lands-waters-threat/ (accessed: 5 December 2018).

Foundation for Democracy and Sustainable Development (2009) 'The relationship between democracy and sustainable development' available at www.fdsd.org/the-challenge/ (accessed: 4 March 2017).

Foundation for Democracy and Sustainable Development (2011) 'British public opinion on the needs of 'future generations'' available at http://www.fdsd.org/publications/british-public-opinion-on-the-needs-of-future-generations/ (accessed: 5 December 2018).

Foundation for Democracy and Sustainable Development (2013) 'The challenge'available at www.fdsd.org/ (accessed: 11 August 2017).

Foundation for Democracy and Sustainable Development (2017) 'Political institutions and policymaking' available at http://www.fdsd.org/ways-forward/political-institutions-and-policy-making/ (accessed: 5 December 2018).

Foundation for the Rights of Future Generations (2003) 'Generational justice – theme of the new century?' Presentation at the workshop "Rights of Future Generations" at Central European University, Budapest, Hungary, 31 January.

Franco, Zeno & Zimbardo, Philip (2006) 'The banality of heroism', *Greater Good*, available at https://greatergood.berkeley.edu/article/item/the_banality_of_heroism/ (accessed: 12 August 2017).

Friends of the Earth (2011) 'Report details world bank's fossil fuel binge' available at https://foe.org/news/2011-04-report-details-world-banks-fossil-fuel-binge/ (accessed: 23 June 2018).

Friends of the Earth Europe (2009) 'Overconsumption? Our use of the world's natural resources' (Vienna, Brussels) available at www.foeeurope.org/publications/2009/Overconsumption_Sep09.pdf (accessed: 7 March 2017).

Future Justice (2016) 'What is future justice?' available at www.futurejustice.org/our-work/ (accessed: 11 August 2017).

Gamson, William (1988) 'The 1987 distinguished lecture: A constructionist approach to mass media and public opinion', *Symbolic Interaction* 11(2): 161–174.

Gamson, William (1992) *Talking Politics*, Cambridge: Cambridge University Press.

Gillot-Assayag, Laure (2016) 'Review of Axel Gosseries & Lukas H. Meyer (eds.) intergenerational justice', *Intergenerational Justice Review*, 1: 32–34.

Global Justice Ecology Project (2016) 'Issues' available at http://globaljusticeecology.org/ (accessed: 11 August 2017).

Goffman, Ervung (1974) *Frame Analysis: An Essay on the Organization of Experience*, London: Harper & Row.

Göpel, Maja (2011) 'Director future justice, world future council. "Guarding Our Future: How to Protect Future Generations"' available at www.thesolutionsjournal.com/node/821 (accessed: 23 July 2016).

Graham, Smith (2015) 'The democratic case for an office for future generations', *Foundation for Democracy and Sustainable Development*, available at http://www.fdsd. org/publications/the-democratic-case-for-an-office-for-future-generations-in-progress/ (accessed: 5 March 2017).

Greenpeace (2017) 'Trump's Latest Executive Order Is an All-Out Attack on Clean Energy' (March 28, 2017) available at www.greenpeace.org/usa/trumps-latest-executive-order-is-an-all-out-attack-on-clean-energy/ (accessed: 28 December 2018).

HelloCoolWorld (2017) 'Occupy your future' available at https://hellocoolworld.com/ (accessed 5 December 2018).

Hill, Damien (2015) 'Bolivia opens up national parks to oil and gas firms', *The Guardian* (5 June) available at www.theguardian.com/environment/andes-to-the-amazon/2015/ jun/05/ (accessed: 4 March 2017).

Hösle, Vittorio (2003) 'Dimensionen der ökologischen Krise – Wege in eine generationengerechte Welt', in Stiftung für die Rechte zukünftiger Generationen (Hrsg.) *Handbuch Generationengerechtigkeit*, München: ökom: 125–150.

Indian Country (2013) 'Fight the war! 13-year-old calls on his generation to save the world' available at http://indiancountrytodaymedianetwork.com/2013/12/30/fight-war-13-year-old-calls-his-generation-save-world-152888 (accessed: 23 July 2016).

Intergenerational Foundation (2016) 'The IF European intergenerational fairness index 2016' available at www.if.org.uk/research/ (accessed: 11 August 2017).

International Covenant on Civil and Political Rights (1966) available at https://www.ohchr. org/Documents/ProfessionalInterest/ccpr.pdf (accessed: 5 December 2018).

International Covenant on Economic, Social and Cultural Rights (1966) available at https:// www.ohchr.org/Documents/ProfessionalInterest/cescr.pdf (accessed: 5 December 2018).

Karpf, David (2010) 'Online political mobilization from the advocacy group's perspective: Looking beyond clicktivism', *Policy & Internet*, 2(4) (December): 7–41. doi: 10.2202/1944-2866.1098.

Klandermans, Bert (1984) 'Mobilization and participation: Social-psychological expansions in resource mobilization theory', *American Sociological Review*, 49(5): 583–600.

Martinez, Xiuhtezcatl (2016) 'Earth guardians youth director, earth guardians' available at www.earthguardians.org/xiuhtezcatl/ (accessed: 5 December 2018).

McKinnon, Catriona (2012) *Climate Change and Future Justice*, London and New York: Routledge.

Movement Generation (2011) 'Justice & ecology project' available at https://movement generation.org/2011/10/ (accessed: 23 June 2018).

Occupy Wall Street (2011a) 'Occupy wall street goes home' available at occupywallst.org/ article/occupy-wall-street-goes-home/ (accessed: 23 June 2018).

Occupy Wall Street (2011b) 'The People versus the 1% Dominance of Wall Street Profi-teers' available at webcache.googleusercontent.com/search?q=cache:0hK7anpHQScJ: https://en.gegenstandpunkt.com/articles/occupy-wall-street-people-versus-1-dominance-wall-street-profiteers+&cd=1&hl=en&ct=clnk&gl=ie (accessed: 6 December 2018).

Our Children's Trust (2016) 'Youth file lawsuit against Norwegian government over Arctic oil' available at https://static1.squarespace.com/static/571d109b04426270152febe0/t/5807bba8579fb39e1dd6639f/1476901803100/NorwayWritSummons.pdf (accessed: 23 March 2016).

Our Children's Trust (2017) 'Legal actions' available at www.ourchildrenstrust.org/ (accessed: 11 August 2017).

Our Generation, Our Choice (2015) 'Youth draw connections between race, environ-ment and immigration' available at http://webcache.googleusercontent.com/search?q=cache:wC4oAoOMSSMJ:www.pubtheo.com/page.asp%3Fpid%3D2047+&cd=1&hl=en&ct=clnk&gl=ie (accessed:6 December).

Rawls, John (1999) 'Justice as fairness: Political not metaphysical', in Freeman, Samuel (ed.) *John Rawls: Collected Papers*, Cambridge, MA: Harvard University Press.

Reiman, Jeffrey (2007) 'Being fair to future people: The non-identity problem in the Origi-nal Position', *Philosophy and Public Affairs*, 35: 69–92.

Rein, Martin (2012) 'Frame reflective policy discourse', in Peter Wagner, Weiss, Carol Hirschon, Wittrock, Björn & Wollman, Hellmut (eds.) *Social Sciences and Modern States: National Experiences and Theoretical Crossroads*, Cambridge: Cambridge Uni-versity Press: 262–289.

Scherman, Andres, Arriagada, Arturo & Valenzuela, Sebastian (2014) 'Student and envi-ronmental protests in Chile: The role of social media', *Politics*, 35(2): 151–171.

Snow, David (2013) 'Framing and social movements', available at https://doi.org/10.1002/978040674871.wbespm343 (accessed: 26 June 2018).

Snow, David & Benford, Robert D. (1988) 'Ideology, frame resonance and collective mobilization', in Klandermans, Bert, Kriesi, Hanspeteer & Tarrow, Sidney (eds.) *From Structure to Action*, Greenwich, CT: JAI Press.

Snow, David & Benford, Robert D. (2000) 'Framing processes and social movements: An overview and assessment', *Annual review of Sociology*, 26: 611–639.

Strydom, Piet (2011) *Contemporary Critical Theory and Methodology*, Oxon: Routledge.

Touraine, Alain (1977) *The Self-Production of Society*, Chicago: University of Chicago Press.

Transition United States (2017) 'Why transition?' available at www.transitionus.org/ (accessed: 11 August 2017).

Tremmel, Joerg C. (2014) *A Theory of Intergenerational Justice*, Oxon: Routledge.

UNICEF UK (2013) 'Climate change: Children's challenge' available at www.unicef.org.uk/publications/climate-change-report-jon-snow-2013/ (accessed: 13 August 2017).

United Nations (2005) 'In larger freedom: Towards development, security and human rights for all' available at www.un.org/en/events/pastevents/pdfs/larger_freedom_exec_summary.pdf (accessed: 11 August 2017).

United Nations Convention on the Rights of the Child (1989) available at www.ohchr.org/Documents/ProfessionalInterest/crc.pdf (accessed: 23 June 2018).

United Nations Environment Programme (2016) 'UNEP frontiers 2016 report: Emerging issues of environmental concern' available at http://wedocs.unep.org/handle/20.500.11822/7664 (accessed: 6 December 2018).

United Nations Framework Convention on Climate Change (1992) available at https://unfccc.int/resource/docs/convkp/conveng.pdf (accessed: 5 December 2018).

United Nations Framework Convention on Climate Change (2015) 'Young and future generations day' (3 December) available at http://unfccc.int/cooperation_support/educa tion_outreach/overview/items/9191.php (accessed: 13 August 2017).

United Nations Human Rights Council (2014) (A/HRC/RES/26/9) available at www.ihrb.org/pdf/G1408252.pdf (accessed: 23 June 2018).

United Nations Kyoto Protocol (1997) available at https://unfccc.int/resource/docs/con vkp/kpeng.pdf (accessed: 5 December 2018).

United Nations Office of the High Commissioner for Human Rights (2018) 'Working group on the issue of human rights and business' available at www.ohchr.org/EN/Issues/Business/Pages/WGHRandtransnationalcorporationsandotherbusiness.aspx (accessed: 26 June 2018).

Van Dijk, Tuen (1977) *Text and Context: Explorations in the Semantics and Pragmatics of Discourse*, London: Longman.

Westall, Andrea (2015) 'The relationship between democracy and sustainable development', *Foundation for Democracy and Sustainable Development*, available at www.fdsd.org/site/wp-content/uploads/2015/06/The-relationship-between-democracy-and-sustainable-development.pdf (accessed: 5 March 2017).

World Future Council (2013) 'Crimes against future generations' available at www.world futurecouncil.org/crimes-against-future-generations/ (accessed: 12 July 2018).

World Future Council (2016) 'Working together to end violations of the human rights of present and future generations by transnational corporations', *Joint Statement* (15 March) available at www.worldfuturecouncil.org/working-together-end-violations-human-rights-present-future-generations-transnational-corporations/ (accessed: 21 July 2016).

World Future Council (2017) 'Global policy action plan' available at www.worldfuture council.org/gpact/ (accessed: 13 November 2017).

World Resources Institute (2014) 'The history of carbon dioxide emissions' available at www.wri.org/blog/2014/05/history-carbon-dioxide-emissions (accessed: 25 June 2018).

Yong, Jung Cho, Shahis, Waleed, Torriente, Devontae & Blazevic, Sara (2015) 'Here's why we're committing civil disobedience: Millennials can no longer be silent about our broken system', *Our Generation, Our Choice Action Team*, (November) available at https://www.salon.com/2015/11/02/heres_why_were_committing_civil_disobedience_millennials_can_no_longer_be_silent_about_our_broken_system/ (accessed: 10 August 2017).

Young, Iris Marion (1997) 'Asymmetrical reciprocity: On moral respect, wonder and enlarged thought', *Constellations*, 3(3): 340–363.

Chapter 3

Are future peoples the bearers of present rights?

Introduction

As it continues to gain societal momentum, future justice campaigners' critique of global capitalism and its ecologically destructive effects opens up new spaces of argumentation in the wider public sphere where differing interpretive positions on the justice dimensions of climate change are explored. Issues of inequality and responsibility are subject to more rigorous forms of validity testing (Habermas, 1984), assessed against truth claims regarding the empirical reality of deepening climate change, rightness claims on the treatment we owe one another and future others, ideas of the good society, as well as optimal means of achieving different societal goals (e.g., sustainable development, conservation, peace, poverty eradication, etc.). The following analysis considers how such processes of reflective public communication unfold today around the question of intergenerational justice – in particular, whether current tendencies to discount the long-term costs of excessive rates of resource depletion can be defended as 'just' or, indeed, ought to be considered 'irresponsible'.

It looks at the type of justifications actors bring to bear on current environmental practices, noting how, for some, present excessive levels of natural resource consumption are a form of 'privileged irresponsibility' (Tronto, 1993, 2013) in that such deeds are not performed in ignorance of their detrimental impact but with significant awareness of their effect on future supplies. For others, however, current resource practices are just and remain an inalienable right of all states' communities, one essential to the exercise of rights to self-determination and development. Whilst divisions between these two interpretive positions continue to deepen, where there is agreement on the power wielded by living generations to save or dispose of natural or material resources and to secure or, indeed, neglect institutions in ways that unborn or younger generations do not exercise power (Tremmel, 2014: 59). Existing legislation encourages natural and cultural heritage to be transmitted to future generations in a manner that is fair and not unreasonably compromised by the products of our living (e.g., excessive carbon pollution, armed conflict, war, etc.). For some, however, there are clear limits to such obligations. Setting aside a portion of scarce resources for the benefit of hypothetical peoples of the future, skeptics argue, makes neither practical nor moral sense, especially when the number of chronically undernourished people in the world is rising steadily (one in every nine people today, according to figures produced by the Food and Agriculture Organization, 2016) and more than 40 percent of the present global population is negatively affected by water scarcity (United Nations, 2017). If justice requires that we address the suffering of those most affected by climate change, critics argue, then the claims of the presently disadvantaged would seem to be more compelling.

The discussion that follows lays out the main arguments presented by exponents of each of these interpretive positions, starting with those firmly against assigning an entitlement status to future generations. It then considers how deteriorating climate conditions, together with growing scientific evidence of long-term deficiencies, are strengthening institutional support for a human rights approach to climate change being explored intergenerationally. Whilst there is broad agreement amongst actors as to the immanent nature of ecological threat today, disagreement arises over how society ought to respond to this situation. The discussion here draws attention to the type of interpretive repertoires competing actors bring to bear on these issues and, more generally, debate on where the boundaries of the just society reside. It employs Wetherell and Potter's (1988) concept of interpretative repertoire as a basic analytic unit to explain the type of socially and culturally familiar arguments actors utilize to develop a credible stance on the notion of intergenerational justice. The assumption is that these actors do not merely offer 'unvarnished' accounts of why the subject of justice must be defined in a specific manner in light of new ecological realities but rather in a way that reflects underlying dimensions of their ideological position and broader worldview (Antaki, 2009: 6).

The case against ascribing present rights to future generations

For opponents, the main objection to assigning legal entitlements, including rights, to future generations is the latter's potential as opposed to real existence (i.e., not yet born). While we may feel certain obligations towards future humanity, such as taking account of the more long-term effects of current environmental, economic and social practices, this does not imply that future persons are presently the bearer of rights or other legal entitlements. Amongst those advocating this viewpoint are Beckerman (1999: 71) and Steiner (1983: 6, 154, 259), both of whom question whether a correspondence can actually be established between legal entitlements that exist now and persons who may or may not exist in the future (depending on circumstances). For both of these thinkers, 'the non existence challenge' proves too big an obstacle to ever make a rights entitlement argument viable in this instance (see also Gosseries, 2008: 453, 2015: 492). How, Steiner (1983: 54) asks, can rights to finite natural resource reserves be enforced if the bearer of these rights is not in a position to exercise them? Having the power to enforce rights to a clean atmosphere, for instance, presupposes that the rights holder and obligation bearer coexist. A certain minimum degree of correlativity, they argue, must prevail between relevant parties if rights are to be actualized in a manner that is meaningful (O'Neill, 2010: 67).

The 'likely future existence' of humans is thought not to offer sufficiently solid grounds for granting a rights persona to unborn peoples, especially as the conditions necessary to sustain healthy living standards deteriorate further (increased levels of drought, prolonged exposure to infectious disease, contamination of food and water supplies, global warming, etc.; see *The Lancet*, 2017). Certainly, as global temperatures continue to rise, the expectation is that prospects for life on this planet will decline, particularly if greater efforts are not made to steer the current energy system onto a safer development path. Present global trends in emissions rates (*National Geographic*, 2017), energy policy (ongoing investment in fossil fuels) and population growth do not suggest for some that commitments to keep temperatures below a 2°C threshold will hold. If agreed safe limits are surpassed, the chemical, physical and biological processes that modulate the Earth's functioning will begin to break down, at which point the capacity of human populations to adapt will start to decline (see, for example, B., Barros et al., 2014; Cropwatch, 2016). Do such future dystopias offer support to the argument of Steiner and Beckerman against assigning rights entitlements to future generations?

If environmental conditions eventually do not support human flourishing, can we meaningfully claim that nonexistent future peoples have been harmed by our failure to preserve a safe planet? Further, if future peoples, by virtue of their nonexistence, cannot be harmed, how can we legitimately claim they have rights? The right and wrong of current resource depleting practices depends almost entirely, according to this perspective, on the consequences of these actions for living generations, not for those not yet born. Climate change undercuts the rights of the living to health, food, water and, especially for some (e.g., small island states), the right to self-determination. Once we begin to talk about these rights, we assume a framework of action in which the performance of legal obligations protecting these rights can be realized. But how can we fulfill obligations to peoples who may never exist? Whatever rights future generations may have in the future, they do not possess these same rights now (Beckerman & Pasek, 2001). At most, humanity is said to bear a moral responsibility to take account of the interests of future generations when shaping the contours of the just society. Obligations here include the need to bequeath to future generations basic liberties and ensure a compassionate treatment of nature (Beckerman, 1999: 86–87). However, in no way are these obligations said to offer a justification for the establishment of a series of noncorrelative rights and duties between specified present and unspecified future peoples. Neither should they be considered a constraint on current standards of living or the pursuit of further economic growth. Also advocating this position, the World Bank in a 2011 report entitled 'Human Rights and Climate Change' notes how 'explicit human rights arguments have yet to gain traction to any appreciable extent within climate change negotiations', especially those referencing future generations. This, it believes, is explained by the absence of any direct support in international law for a rights approach to intergenerational justice:

> The international human rights framework, on its face, appears not to accommodate easily the interests of future generations. The human rights legal framework appears somewhat reactive in its design, geared more to redressing part of imminent harms than the speculative business of scientific predictions pertinent to climate change.
>
> (2011: 47)

The sheer 'speculative' nature of scientific assessments of future ecological conditions, it argues, should discourage any efforts to extend a human rights framework outward to include future peoples. A human rights approach is even said to risk 'overloading an already fragile climate change agenda', chiefly because human rights are a 'source of mistrust' amongst developing states concerned they will be used 'as a way of either preventing their development' or imposing 'conditionalities' on their eligibility for climate change adaptation funds (ibid.: 10). Amongst more industrialized states, also, the World Bank detects a fear that a stronger official recognition of linkages between climate change and human rights violations will bolster the case for further unwanted extra-territorial legal regulations. Causing particular concern is the 'collective or self-standing right to a safe and secure environment' and its use as a 'political or legal weapon against them' (ibid.: 10). Certainly, such concerns were expressed by the United States in its submission to the UN Office of the High Commissioner for Human Rights in 2009 on the relationship between climate change and human rights when it denied the existence of a right to 'a safe environment' under international law and, by extension, the legitimacy of any efforts to promote rights arguments on this basis (see Observations by the United States of America on the Relationship between Climate Change and Human Rights, 2009). Climate change, it added, is 'one of many natural and social phenomena that may affect the enjoyment of human rights' and, therefore, cannot be singled out as 'the cause' of human rights violations, particularly those arising internationally (ibid.: 4). Restricting resource rights eligibility to living 'legitimate' claimants, particularly those with a legal contractual right to precious reserves of minerals, oil, gas, seeds, forests, arable lands and the like, and striking 'a balance' between environmental harms and the benefits of the activities causing it are asserted instead as primary concerns, as is the need to protect the energy, water and development needs of the present (see, for example, World Bank, 2017: 1–2).

According to this perspective, rights eligibility cannot be extended to future peoples because, amongst other things, they are not yet the bearers of specific properties. In other words, to 'have rights' is to possess properties to which one can claim a legitimate right. What is proposed instead is a weak consideration of the interests of future peoples but not their rights. In making this argument, the assumption would be that one can prove the absence of any added value in ascribing rights to future peoples at this point in time. The assumption also would be that one can demonstrate how current rates of depletion of fossil fuels, forests, fish stocks, arable lands, in gravely affecting future supplies, affect only the interests of future generations but not necessarily their rights to health (e.g., International Covenant on Economic, Social and Cultural Rights, 1966, Article 12), development (International Covenant on Civil and Political Rights, 1966, Article 1), a safe environment (constitutionally grounded) or freedom from want (United Nations Universal Declaration on Human Rights, 1948). As the science of climate change grows more precise, so too does our knowledge of its future impact on the ecological circumstances of generations to come. If current resource depletion practices are not performed in ignorance of their long-term effects but rather with significant awareness, do they not correlate directly with negligence on our part? For those opposed to assigning resource entitlements to future others or being heavily bound by duties to the same, the answer is unquestionably 'no'. The fact that when we speak of future generations being harmed by current environmental practices, we speak of no one in particular (unspecified persons) but a general category of humanity (Reiman, 2007: 78) makes the appeal to rights inappropriate for exponents of this interpretive position. Others, however, disagree, arguing that when a group of experts representing a broad range of interests deliberate on which principles of justice ought to determine how we approach a given issue, they do so largely behind 'a veil of ignorance' (Rawls, 1999: 15–19), that is, without knowledge of the specific circumstances or characteristics of relevant parties (e.g., race, gender, age or social position). Similarly, principles of justice can be extended to unspecified future others (e.g., Reiman, 2007: 79) on the basis of the same reasoning.

The case in favor of ascribing present rights to future generations

For legal experts and human rights campaign groups, the World Bank's reading of human rights eligibility is objectionable for a number of reasons. First, it signals what they believe to be a deliberate misinterpretation of the essential nature of human rights (not just a matter of entitlement but also duties to others). Rights, they argue, cannot be equated with ownership of the world's remaining lands, freshwater reserves, seeds, minerals and other essential resources in the manner implied by this interpretive position, where emphasis is overwhelmingly placed on the needs of the present. The basis of these actors' critique is both practical and ideological.

Ideologically, a private property interpretation of entitlement is countered with a common ownership one. Here, the focus is on natural resource legacies and the 'planetary responsibilities' (Brown-Weiss, 1989: 2) of all to hold them in trust for generations to come. There is no question in this instance of excluding others (include future generations) from gaining access to essential resources on the basis of claims of 'having been first!' (Waldron, 2003: 82). In practical terms, also, a private property model of entitlement is said to be of limited value given the growing complexity and the nonsymmetric nature of the relationship between sources of climate harm (geographically and temporally dispersed) and their victims (present *and* future generations). Whilst in some instances, there is still a clear, identifiable relationship between rights holders and obligation bearers (those who coexist in time), in many other instances there is not (e.g., cumulative harms experienced across generations).

Many of the basic insights of the science of climate change, including those on the protracted nature of ecological agency, run contrary to those of traditional approaches to resource justice (e.g., the lack of direct evidence of ecological harm, the non-copresence of obligation bearers and rights recipients, the assignment of blame for a discrete act of pollution to a particular party in the interests of restoring normality, etc.). What are commonly expressed in relation to newer cognitive understandings of Anthropocene worlds are not distinctions between but rather connections that bind present and future generations together. In this context, resource needs, including those arising from persisting poverty and growing natural resource scarcity worldwide, are as much a matter of intergenerational as intragenerational equity, given the tendency for poverty and deprivation to be transmitted from parent to child and declining rates of social mobility (UN General Assembly, 2013; Philip Alston, UN Special Rapporteur on Extreme Poverty and Human Rights, December 2017).

UN Special Rapporteur on Extreme Poverty and Human Rights, Philip Alston (2015) accuses the World Bank of trying to create a 'human rights free zone' and portrays the capitalist order it supports as a hazard to the basic needs of all of humanity. Equally critical, Human Rights Watch (2015) describes the World Bank's second draft of its safeguard policies for nearly 12,000 high-risk projects in over 170 countries (World Bank, 2015) as a 'dangerous rollback in environmental [and] social protections'.[1] Instead of providing 'heightened protection and vigilance' at this crucial point in our ecological history (in terms of long-term sustainability), such a policy is thought to weaken commitments to international human rights standards and increase the 'discretionary nature' of environmental protections. The Centre for International Environmental Law further questions how more voluntary regulatory arrangements could be seen as a preferable means of protecting the welfare of vulnerable communities, present and future (CIEL, 2016). It, together with a growing number of legal agencies, human rights organizations, and climate justice campaigners, draw on the 'evidentiary' knowledge of science (CIEL, 2017), as much as legal arguments to support a position of opposition to attempts to restrict justice considerations (including human rights protections) to the needs and dominant capitalist interests of the present.

Harmful acts of resource destruction, they argue, may occur at a distance and during a time period when the primary victim is not yet born, but this does not detract from duties to ensure that 'due regard' is shown towards 'the human rights and fundamental freedoms' of such peoples (e.g., see UN Declaration on the Responsibilities of Present Generations towards Future Generations, 1997, Article 2), including a right to legal representation. We may be uncertain at this point as to what the good life will mean for peoples of the future or how precisely populations will adapt to living under harsh ecological conditions. However, we do know two basic facts: first, what basic resources populations need to survive and, second, given their finite and precious nature, our responsibilities to ensure that such resources are protected. Since future generations are helpless to change what is forced upon them (e.g., compromised ecosystems), it is incumbent on living peoples to impose lesser risks than are currently being created through drilling, ocean trawling, fracking and similarly destructive practices (Greenpeace, 2016). Since these practices directly impact upon the health of the planet, they inevitably will also impact upon the health of future generations and restrict the range of choices available to them (e.g., choices as to where to live [habitable lands], what crops will grow, the distribution of freshwater supplies, illnesses and diseases caused by cumulative atmospheric pollution, etc.). In this sense, current pollution practices impact upon the capacities of future peoples to enjoy legally protected rights to heath, development, freedom and security. It is essential, therefore, that present generations do not act as if current rates of resource depletion do not matter.

The fact that future peoples cannot impose constraints on current rates of resource consumption does not take away from obligations to respect their rights to a portion of these resources (see, for instance, Constitution of Japan, Article 11, Chapter III, 1946). Article 4 of the UNESCO Declaration on the Responsibilities of the Present Generations towards Future Generations (1997) refers to obligations 'to bequeath to future generations an Earth which will not one day be irreversibly damaged by human activity'. Article 5 stresses the need to preserve 'for future generations resources necessary for sustaining human life and development'. Because persons belonging to future generations are recognized as having of this moment certain rights by the UNESCO (1994) Universal Declaration of the Rights of Future Generations and the Rio Declaration (1997, Principle 3), both offer clarification as to why the welfare *and rights* of future peoples must be legally protected. Article 2 of the Convention on Biological Diversity (1992) highlights the importance of exploiting biological diversity in a 'sustainable' manner and at a rate that does not lend itself to long-term decline. The potential of ecosystems, it adds, must be maintained 'to meet the needs and aspirations of present and future generations'. Such legislation prohibits discriminatory denials of access to basic resources through severe environmental harm and pollution. Similarly, Article 2(5) of the Convention on the Protection and Use of Trans-Boundary Watercourses and International Lakes (1992) specifies how 'water resources shall be managed so that the needs of the present generation are met without compromising the ability of future generations to meet their own needs'.

Of relevance here are both the individual rights of 'identifiable' living peoples to a sustainable future, as well as the collective rights of future generations whose identity may not be clearly determinable at this point in time but whose need for basic, life-enabling resources is. Freeman (1995: 30) in his reading of Raz's (1986: 166) interest-based concept of rights, suggests that A (e.g., future generations) has a right to x (natural resources as public goods) only if the interest A has in having x is a sufficient reason for imposing a duty on B (e.g., living generations) to protect A's right to x. Given that the resources in question are basic to survival (e.g., clean water, air and lands), A's interest in x in this instance is incontestable. Because the preservation of safe and healthy living conditions is essential, sufficient reason is present to impose strict duties on living generations to ensure that ecological conditions continue to support planetary life into the future. The interests of no single member of future generations to such resources are sufficient to justify holding present resource users accountable for duties of care and legal responsibility (as philosophers Beckerman and Steiner highlight). However, the fundamental interests of all together are sufficient. What is required, therefore, is a broader approach to intergenerational justice, one that sees environmental security as a public good held across generations (not the sole preserve of the living or the wealthy). The essence of this approach is the normative protection of *all generations'* ecological inheritance, basic well-being, and ongoing capacity to be self-determining (to enable sufficient economic, social and cultural development).[2]

In spite of opposition, grounding human rights intergenerationally and formulating justice in deeper temporal terms could be said to be present in law already (albeit in a largely implicit form). References to future generations and a responsible use of natural resources are evident across a range of legal instruments, some of which have been discussed earlier, but also in a significant number of state constitutions. In the case of the latter, references tend to be either general provisions for the protection of future peoples or more specific references to the natural environment (see Constitution of the Plurinational State of Bolivia, 2009). The constitution of Norway, for instance, recognizes the right of 'every person' to an environment that is conducive to health and emphasizes that 'this right will be safeguarded for future generations as well' (see Constitution of Norway, Article 112). Others focus on the responsibilities of the state to act as a guardian of the resource commons. Article 20a of the German constitution, for example, defines the role of the state as protecting 'the natural living conditions' of all, including 'future generations' (Basic Law for the Federal Republic of Germany, 2014). Similarly, the constitutions of the Netherlands and Switzerland include a government mandate to fulfill responsibilities towards future peoples. Existing legal instruments, therefore, offer an important institutional basis on which to argue in favor of a stronger, more detailed legal interpretation of intergenerational justice. However, there is one other interpretive tool that proves indispensable, and that is the research findings of climate scientists. The 'evidentiary basis' (CIEL, 2017) of this research and the unparalleled degree of consensus that has emerged amongst science communities in recent years as to the nature of ecological threats and their span of affectivity provides a new level of institutional validity to this interpretive position.

The contribution of science to the debate on intergenerational justice

Using climate models of increasing complexity, scientific research today documents precisely how and why global temperatures and atmospheric pollution are rising faster than previously assumed (see IPCC Fifth Assessment Report, 2013).[3] So detailed are these research findings, they prompt a fundamental reconsideration of the way human–environmental interactions have traditionally been conceptualized. Industrial-scale energy flows from fossil fuel carbon, in particular, are shown to be transforming the Earth's atmosphere, oceans, biosphere and nutrient cycles (Raupach & Canadell, 2010) at a pace that can no longer be described as within the boundaries of a 'safe operating space' for humanity and, indeed, the rest of nature.[4] Increasing knowledge of 'the deep time scales' of human interference with the natural cycles of the planetary system (Chakrabarty, 2017: 42) now triggers a fundamental reconsideration of the long-term justice dimensions of such practices. With greater understanding of how environmental harms unfold across multiple time frames (and implicate many generations of victims), the realization is that those who contribute to escalating rates of GHG pollution and those who endure its worst effects do not necessarily coexist within the same time period. Carbon dioxide concentrations in the atmosphere, for instance, cause global warming and global mean sea levels (GMSL) to rise for centuries (IPCC Fifth Assessment Report, 2013) and trigger effects such as ocean acidification, changes in soil abiotic conditions, reduced fertility and prolonged drought conditions over time. With more precise scientific evidence of how these harms unfold over many years comes the realization that presence can no longer be considered a prerequisite for the imposition of harm.

The 'deep time' of ongoing eco-destruction necessitates a newer understanding of ecological agency, one that critically reevaluates relevant contexts for the application of principles of justice and understands, on the basis of latest scientific evidence, how human existence is intricately intertwined with multiple others located across space and time. Chakrabarty (2017) notes how this knowledge is institutionally registered. In the immediate aftermath of the publication of the Fourth Assessment Report of the Inter-Governmental Panel on Climate Change (IPCC Fourth Assessment Report, 2007), for instance, attention began to shift more substantially to the justice implications of human-made planet vulnerabilities and risks. In particular, the ramifications of a growing lack of symmetry between documented sources of climate harm (geographically and temporally dispersed) and their victims (present *and* future planetary life). Newer scientific evidence did not offer unequivocal support for the historical coexistence of climate harm doers and their primary victims. Instead, it provided a graphic illustration of how pollution practices outrun their original point of initiation and counteract what many regard as their original intention (i.e., to advance human progress).

Each act of ecological destruction, in fact, could now be shown to be part of a much larger material circuitry of agency that completes itself only gradually in terms of its effects on the wider planetary system. The type of arguments put forward in defense of a restriction of rights eligibility to the present no longer seemed entirely convincing in light of newly emerging empirical truths. As a matter of necessity, models of justice have become the subject of greater societal scrutiny. Taking more than is required to live healthy sustainable lives comes to be seen by increasing numbers as an injustice. While this is not necessarily a 'new' argument, the extent to which publics today are willing to criticize government policy and take action in an effort to control pollution levels is. It was only a matter of time before understandings of human rights eligibility would be officially reexamined and notions of human rights agency redefined.[5]

Tendencies to accord privilege to the resource rights of presently living generations are, as a consequence, subject to more regular dispute today. So too are the underlining premises of a narrowly defined liability approach to environmental justice (where emphasis is placed on direct evidence of harm and clear victim and perpetrator identities), in particular, the nonreflexive assumptions of presentism embedded in this approach (Thompson, 2010).[6] Today we find ourselves confronted with a series of problems for which a broader perspective is urgently required. No obvious solutions present themselves for climate-induced migration (e.g., no legal protections for its victims), no formal settlement arrangements have been established for communities threatened by rising sea levels or desertification, no comprehensive legal arrangements have been forged to address likely future conflict linked to water shortages or the sharing of transboundary river sources, to name but a few, chiefly because the source of 'wrongdoing' in each case is not straightforward. Efforts to restrict deliberation on wrongdoing to evidence of direct harm prove part of the problem (hence, the decision of the International Criminal Court [2016] recently to include 'indirect victims' as relevant categories in its criminal investigations). Dispersed acts of climate harm, in terms of their effects on vulnerable peoples, do not reach an end point that is readily identifiable (an implicit assumption of a narrowly defined liability approach to justice). Predilections towards a linear logic now have to make room for the discovery of nonlinear relations of harm. Environmental cause and effect cannot be conceived exclusively in terms of double contingent ego-alter relations (e.g., clear perpetrators and victims) but rather must be formulated in terms of more complex ego-alter-other relations where third parties (e.g., generations unborn, wider ecosystems, etc.) are taken into account (i.e., as 'indirect victims').

Scientific research on climate change offers much in the way of food for thought. However, there is still a worrying lack of evidence in the wider policy field of efforts to respond to these new cognitive insights and engage in forward planning of a type that targets the deeper, more structurally embedded nature of ecological destruction. Without a more long-range perspectivism on duties, rights and responsibilities, critics argue, climate problems are likely to regress even further. Above all, the type of ontologies of presence (i.e., the cultural values and meanings that guide dominant interpretations of what is present and relevant to the world) that influence legal and political, as well as economic approaches to environmental justice to date, must be critically reconsidered. Knowledge of the plural temporalities of threats to planetary well-being necessitates a deeper justice framework.

No one makes this point more cogently today than concerned citizens who insist that poor regulatory performance in pollution control be addressed as a matter of urgency. This means examining the procedural, as much as interpretive dimensions of intergenerational justice with greater legal precision. Intergenerational justice must be a matter of assignable, claimable, institutionalizable and enforceable rights that hold ground across time. The fact that the interests of present and future generations are referenced by a range of existing legal instruments (including state constitutions, international declarations and conventions) is, arguably, in itself not sufficient to ensure that the necessary changes occur in procedures of justice or result in better decision-making outcomes (e.g., the implementation of a series of corrective measures, the banning of certain pollutants or the immediate revision of nonrenewable energy targets downward in recognition of the detrimental effects on global warming of burning further fossil fuels). An underlining bias towards the present as the most relevant context of justice remains intact.[7] A deeper justice framework, alternatively, requires a type of 'democratic futurism' where manageable thresholds of CO_2 emissions are planned for with the future firmly in mind and a critical loss of specific possibilities for justice, freedom and human flourishing consciously avoided.[8]

It is the responsibility of all states to ensure 'the identification, protection, conservation, presentation and transmission to future generations of the cultural and natural heritage' of this world (see Convention Concerning the Protection of the World Cultural and Natural Heritage, 1972, Article 4). Furthermore, the expectation is that states will protect the resource interests of present and future generations in a manner that facilitates the capacities of both to be self-determining and will do so in a way that respects the rights of all state communities to do the same. The Responsibility to Protect (A/RES/63/308), endorsed by the UN General Assembly in July 2009 (UN, 2009), clarifies, amongst other things, each sovereign state's responsibility to protect the peoples within its borders from gross violations of human rights (UN Office of the High Commissioner for Human Rights, 2011). The question being explored at present is whether state-endorsed pollution practices represent violations of the responsibility to protect and uphold human rights law? In principle, when states fail to fulfill such obligations, legal action can be taken against them. While there are no historical precedents to date of cases where the international community has intervened when a state has allowed the natural resource base of its peoples to be depleted to grave levels and, in that, has violated their rights, there is, nonetheless, a growing number of cases where concerned citizens have taken it upon themselves to hold states legally accountable for neglecting their natural resource interests and human rights (e.g., their rights to health, liberty and development) and for failing to safeguard these rights for future generations as well (e.g., *Greenpeace & Nature and Youth v. the State of Norway*, 2017; *Dejusticia v. the State of Colombia*, 2018). In that, law offers publics a mechanism with which to challenge state actions deemed harmful to their welfare. In the following chapter, several of these cases will be explored in detail. Suffice to say for now that legal challenges are an important reminder to states of constitutional obligations to protect the interests and, indeed, entitlements of overlapping generations to collective environmental goods (see United Nations General Assembly, 2013). More generally, legal challenges give us some sense of how 'the social' and 'legal life of rights' (Wilson, 2006; Forst, 2017: 74) continues to evolve, as publics try to make sense of the justice dimensions of deteriorating climate conditions worldwide. In the process, new iterations of ideals of freedom, democracy and right are produced, and states come under pressure once again to invest in a common democratic future (World Commission on Environment and Development, 1987; See, also. United Nations, Back to Our Common Future, 2012).

The type of interpretive work generated through such legal discourse, as well as wider justice campaigns, reaffirms the relevance of human rights and legal responsibilities to an expanding range of climate change issues. This, combined with the increasing tendency for sympathetic court judges to support the constitutional validity of publics' 'climate rights claims', suggests that the type of evaluative frameworks being brought to bear on environmental issues are beginning to change.[9] Cumulatively, these developments hint at new conditions of possibility for a deeper justice framework to emerge, at least in the legal sphere. The settled convictions of traditional state perspectives on pollution practices are subject to more regular criticism (e.g., a narrowly defined liability approach to environmental issues), as are established orders of justification that prioritize the interests and rights of present polluters over all others. This, together with more recent moves, including the launch of the UN's new Environmental Rights Initiative (United Nations News, March 2018), would suggest that a degree of institutional learning is under way.[10] The following chapter looks at the type of reflective learning being generated through legal cases, where the ongoing validity of human rights and constitutional law is actively reconstructed by citizens in light of newer ecological, social and political challenges.

Notes

1 According to a recent report published by the ICIJ (2015), between 2009 and 2013, World Bank Group lenders invested $50 billion in projects graded at the highest level of risk for 'irreversible or unprecedented' environmental impact. That is more than twice as much as that invested in similar projects in the previous five-year period.

2 Self-determination, as a fundamental principle of human rights law, is considered a norm of *jus cogens* (a primary rule), and it has been defended by the International Court of Justice, as having the status of *erga omnes*, that is, a human right 'flowing to all' (Parker, 2000).

3 See also Taylor, Patrick and Holdren, John P. Clouds and the Earth's Radiant Energy System (CERES) 'What happens in the Arctic doesn't stay in the Arctic' (2015), https://ceres.larc.nasa.gov/ (accessed March 2, 2018).

4 See, for example, P. Goodwin, A. Katavouta, Y. Roussenov, M, Foster, L. Gavin, E. J. Rohling & R. Williams (2018) 'Pathways to 1.5 and 2°C warming based on observational and geological constraints', *Nature Geoscience,* 11), 102–107.

5 Report of the Office of the United Nations High Commissioner for Human Rights on the Relationship between Climate Change and Human Rights (2009); United Nations (2013) Report of the UN Secretary General, 'Intergenerational Solidarity and the Needs of Future Generations', available at www.futurejustice.org/wp-content/uploads/2013/10/N1342845.pdf (accessed: 6 December 2018). See, also, Human Rights Council (2016) 'Report of the special rapporteur on the issue of human rights obligations relating to the enjoyment of a safe, clean, healthy and sustainable environment'.

6 Presentism may be described as the tendency to neglect the long-term environmental implications of current decision making (see Thompson, 2010: 18).

7 See D. F. Thompson, 'Representing future generations: political presentism and democratic trusteeship', *Critical Review of International and Political Philosophy*, 13(1) (2010), 18.

8 A number of constructive proposals along these lines have been presented in more recent years, including the introduction of The Well-being of Future Generations (Wales) Act, which obliges public bodies to carry out sustainable development and well-being assessments on their operations, the results of which are published in annual future trends reports to the Welsh Assembly. Another proposal is that made by former UN Secretary General Ban Ki-Moon, in the 2013 report 'Intergenerational Solidarity and the Needs of Future Generations' on the possibility of a new High Commissioner Office for Future Generations being established and short-term thinking in all policy sectors reoriented towards more long-term goals. The explicit aim in this instance is to ensure that the core principles of the UN Charter and those of the UN's sustainable development goals are merged. The legal rationale for the establishment of such an office would stem from existing international legal commitments to the protection of all generations across time (see, for example, the Preamble of the International Covenant on Economic, Social and Cultural Rights [1966], which refers to 'the equal and inalienable rights of all members of the human family').

9 For example, see Superior Court of the State of Washington, Judge Hollis R. Hill, December 19, 2016.

10 The UN's new Environmental Rights Initiative is designed to encourage commercial actors to be more proactive 'champions' of a clean and democratic future. Coinciding with the launch of this new initiative, UN High Commissioner for Human Rights Zeid Ra'ad Al Hussein describes how 'violations of environmental rights have a profound impact on a wide variety of human rights, including rights to life, self-determination, food, water, health, sanitation, housing, cultural, civil and political rights'. Special Rapporteur on Human Rights and the Environment John Knox added that 'the time has come to recognize the formal interdependence of human rights and the environment not only at national but at the UN level also' (see UN News, 6 March 2018).

References

Alston, Philip (2017) 'Statement on visit to the USA', Office of the High Commissioner for Human Rights (15 December) available at www.ohchr.org/EN/NewsEvents/Pages/DisplayNews.aspx?NewsID=22533 (accessed: 10 July 2018).

Alston, Philip (2015) 'Report of the special rapporteur on extreme poverty and human rights to the UN General Assembly' (4 August) available at https://www.ohchr.org/EN/Issues/Poverty/Pages/SRExtremePovertyIndex.aspx (accessed: 14 March 2017).

Antaki, Charles (2009) 'Analysing discourse', in Alasuutari, Pertti, Bickman, Len & Brannen, Julia (eds.) *Handbook of Social Research Methods*, London: Sage.

Beckerman, Wilfred (1999) 'Sustainability and intergenerational justice', in Dobson, Andrew (ed.) *Fairness and Futurity*, Oxford: Oxford University Press.

B., Barros, Vicente R., Mastrandrea, Michael D. et al. (2014) *Climate Change 2014: Impacts, Adaptation, and Vulnerability, Part B: Regional Aspects, Contribution of Working Group II to the Fifth Assessment Report of the Intergovernmental Panel on Climate Change*, Cambridge: Cambridge University Press.

Beckerman, Wilfred & Pasek, Joanna (2001) *Justice, Posterity and the Environment*, Oxford: Oxford University Press.

Brown-Weiss, Edith (1989) *In Fairness to Future Generations*, Tokyo and New York: United Nations University and Transnational Publishers.

Centre for International Environmental Law (CIEL) (2016) 'Glass half full? The state of accountability in development finance' available at www.ciel.org/wp-content/uploads/2016/01/IAM_DEF_WEB.pdf (accessed: 6 December 2018).

Centre for International Environmental Law (CIEL) (2017) 'Smoke and fumes: The legal and evidentiary basis for holding big oil accountable for the climate crisis' available at www.ciel.org/reports/smoke-and-fumes/ (accessed: 9 July 2018).

Chakrabarty, Dipesh (2017) 'The future of the human sciences in the age of humans: A note', *European Journal of Social Theory*, 20(1): 39–43.

Constitution of Japan (1946) available at www.solon.org/Constitutions/Japan/English/english-Constitution.html (accessed: 4 August 2017).

Constitution of Norway (1814) available at www.stortinget.no/en/In-English/About-the-Storting/The-Constitution/ (accessed: 9 July 2018).

Constitution of the Plurinational State of Bolivia (2009) available at www.constitutepro ject.org/constitution/Bolivia_2009.pdf (accessed: 4 August 2017).

Cropwatch (2016) 'Impacts of extreme heat stress and increased soil temperature on plant growth and development' available at http://cropwatch.unl.edu/2016/impacts-extreme-heat-stress-and-increased-soil-temperature-plant-growth-and-development (accessed: 8 August 2016).

Dejusticia (2018) 'In historic ruling, Colombian court protects youth suing the national government for failing to curb deforestation' (5 April,) available at https://www.dejus ticia.org/en/en-fallo-historico-corte-suprema-concede-tutela-de-cambio-climatico-y-generaciones-futuras/ (accessed: 9 July 2018).

Deutscher Bundestag (2014) 'Basic law for the federal republic of Germany' available at www.btg-bestellservice.de/pdf/80201000.pdf (accessed: 4 August 2017).

Field, Christopher B., Barros, Vicente R., Mastrandrea, Michael D. et al. (2014) *Climate Change 2014: Impacts, Adaptation, and Vulnerability, Part B: Regional Aspects, Contribution of Working Group II to the Fifth Assessment Report of the Intergovernmental Panel on Climate Change*, Cambridge: Cambridge University Press.

Food and Agriculture Organization, International Fund for Agricultural Development (2016) *World Food Program, 'The State of Food Insecurity in the World 2015. Strengthening the Enabling Environment for Food Security and Nutrition'*, Rome: FAO, available at http://www.fao.org/3/a-i4646e.pdf (accessed: 19 September 2016).

Forst, Rainer (2017) 'A critical theory of human rights- some groundwork', in Deutscher, P. & Lafont, C. (eds.) *Critical Theory in Critical Times: Transforming the Global Political and Economic Order*, New York: Columbia University Press: 74.

Freeman, Michael (1995) 'Are there collective human rights?' *Political Studies*, XLIII: 25–40.

Gosseries, Axel (2008) 'Future generations' future rights', *The Journal of Political Philosophy*, 16(4): 446–474.

Gosseries, Axel (2015) 'Introduction: Representing future generations?' *Jurisprudence*, 6(3): 492–495.

Greenpeace (2016) 'This far, no further: Protecting the Arctic from destructive trawling' available at www.greenpeace.org/international/Global/international/publications/climate/2016/This-Far-No-Further.pdf (accessed: 12 March 2017).

Greenpeace (2017) 'Greenpeace and nature and youth take the Norwegian Government to the supreme court' (5 February 2018) available at www.greenpeace.org/norway/no/nyheter/2018/Greenpeace-and-Nature-and-Youth-take-the-Norwegian-Government-to-the-Supreme-Court/ (accessed: 9 July 2018).

Greenpeace Norway & Nature and Youth (2017) 'The climate lawsuit against the Norwegian government' available at www.savethearctic.org/en/peoplevsarcticoil/background-documents/ (accessed: 9 July 2018).

Habermas, Jurgen (1984) *The Theory of Communicative Action Vol 1: Reason and the Rationalization of Society*, Boston: Beacon.

Human Rights Council (2016) 'Report of the special rapporteur on the issue of human rights obligations relating to the enjoyment of a safe, clean, healthy and sustainable environment', *General Assembly* (A/HRC/31/52), 1 February 2016.

Human Rights Watch (2015) 'World Bank: Dangerous rollback in environmental and social protections' (4 August) available at www.hrw.org/news/2015/08/04/world-bank-dangerous-rollback-environmental-social-protections (accessed: 1 August 2016).

Inter-governmental Panel on Climate Change (2007) 'IPCC fourth assessment report' available at https://www.ipcc.ch/pdf/assessment-report/ar4/wg2/ar4_wg2_full_report.pdf (accessed: 6 December 2018).

Inter-governmental Panel on Climate Change (2013) 'IPCC fifth assessment report' available at www.ipcc.ch/report/ar5/ (accessed: 9 July 2018).

International Consortium of Investigative Journalists (2015) 'Evicted and abandoned: How the World Bank broke its promise to protect the poor' available at http://www.ipcc.ch/report/ar5/ (accessed: 6 December 2018)

International Covenant on Civil and Political Rights (1966) available at www.ohchr.org/EN/ProfessionalInterest/Pages/CCPR.aspx (accessed: 3 August 2017).

International Covenant on Economic, Social and Cultural Rights (1966) available at www.ohchr.org/EN/ProfessionalInterest/Pages/CESCR.aspx (accessed: 3 August 2017).

International Criminal Court (2016) 'Office of the prosecutor, policy paper on case selection and prioritisation' available at www.icc-cpi.int/itemsDocuments/20160915_OTP-Policy_Case-Selection_Eng.pdf (accessed: 6 December 2018).

The Lancet (2017) 'Estimates and 25-year trends of the global burden of disease attributable to ambient air pollution: An analysis of data from the Global Burden of Diseases Study 2015', 389(10082) (May) available at www.thelancet.com/journals/lancet/article/PIIS0140-6736(17)30505-6/fulltext (accessed: 20 November 2017).

National Geographic (2017) 'Global carbon dioxide emissions are rising again – latest stories (13 November) https://news.nationalgeographic.com/2017/11/climate-change-carbon-emissions-rising-environment/ (accessed: 2 March 2018).

O'Neill, Onora (2010) 'A Kantian approach to transnational justice', in Brown, Garrett W. & Held, David (eds.) *The Cosmopolitan Reader*, Cambridge, MA: Polity Press.

Oxfam (2015) 'World Bank safeguards review phase two' available at https://consultations.worldbank.org/Data/hub/files/world_bank_safeguards_review_oxfam_submission_phase_two_final.pdf (accessed: 6 December 2018).

Parker, Karen (2000) 'Understanding self-determination' available at http://www.guidetoaction.org/parker/selfdet.html (accessed: 6 December 2018).

Raupach, Michael R. & Canadell, Josep G. (2010) 'Carbon and the Anthropocene', *Current Opinion in Environmental Sustainability*, 2: 210–218.

Rawls, John (1999) *A Theory of Justice*, Harvard: Harvard University Press.

Raz, Joseph (1986) *The Morality of Freedom*, Oxford: Clarendon Press.

Reiman, Jeffrey (2007) 'Being fair to future people: The non-identity problem in the original position', *Philosophy & Public Affairs*, 35(1): 69–92.

Steiner, Hillel (1983) 'The rights of future generations', in MacLean, Douglas & Brown, Peter G. (eds.) *Energy and the Future*, Totowa, NJ: Rowman & Allenheld.

Thompson, Dennis (2010) 'Representing future generations: Political presentism and democratic trusteeship', *Critical Review of International and Political Philosophy*, 13(1): 17–37.

Tremmel, Jorg C. (2014) *A Theory of Intergenerational Justice*, New York and London: Routledge.

Tronto, Joan C. (1993) *Moral Boundaries: A Political Argument for an Ethic of Care*, New York: Routledge.

Tronto, Joan C. (2013) *Caring Democracy: Markets, Equality and Justice*, New York: NYU Press.

United Nations (2009) 'Responsibility to protect' available at www.un.org/en/preventgenocide/rwanda/about/bgresponsibility.shtml (accessed: 4 August 2017).

United Nations (2012) '*Back to Our Common Future: Sustainable Development in the 21st Century*' available at https://sustainabledevelopment.un.org/content/documents/UN-DESA_Back_Common_Future_En.pdf (accessed: 2 March 2018).

United Nations (2017) 'Water' available at www.un.org/en/sections/issues-depth/water/ (accessed: 13 March 2017).

United Nations Convention on Biological Diversity (1992) available at www.cbd.int/doc/legal/cbd-en.pdf (accessed: 31 July 2016).

United Nations Convention Concerning the Protection of the World Cultural and Natural Heritage (1972) available at legal.un.org/avl/ha/ccpwcnh/ccpwcnh.html (accessed: 9 July 2018).

United Nations Convention on the Protection and Use of Transboundary Watercourses and International Lakes (1992) available at www.unece.org/fileadmin/DAM/env/water/pdf/watercon.pdf (accessed: 3 August 2017).

United Nations Declaration on the Responsibilities of the Present Generations towards Future Generations (1997) available at http://portal.unesco.org/en/ev.php-URL_ID=13178&URL_DO=DO_TOPIC&URL_SECTION=201.html (accessed: 31 July 2016).

United Nations Educational, Scientific Cultural Organization (UNESCO) (1994) 'Universal declaration of the rights of future generations', available at www.grimee.com/100169Eo.pdf (accessed: 3 August 2017).

United Nations Educational, Scientific Cultural Organization (UNESCO) (1997) 'Declaration on the responsibilities of the present generations towards future generations' available at http://portal.unesco.org/en/ev.php-URL_ID=13178&URL_DO=DO_TOPIC&URL_SECTION=201.html (accessed: 20 August 2017).

United Nations General Assembly (2013) 'UN report on intergenerational solidarity and the needs of future generations' (5 August) available at https://sustainabledevelopment.un.org/content/documents/2006future.pdf (accessed: 9 July 2018).

United Nations News (2018) 'UN launches environmental rights initiative' (6 March) available at https://news.un.org/en/story/2018/03/1004242 (accessed: 9 July 2018).

United Nations Office of the High Commissioner for Human Rights (2009) 'OHCHR study on the relationship between climate change and human rights' available at www.ohchr.org/EN/Issues/HRAndClimateChange/Pages/Study.aspx (accessed: 9 July 2018).

United Nations Office of the High Commissioner for Human Rights (2011) 'Guiding principles on business and human rights for implementing the UN "Protect, Respect and Remedy" Framework' (16 June) available at www.ohchr.org/Documents/Publications/GuidingPrinciplesBusinessHR_EN.pdf (accessed: 4 August 2017).

United Nations Universal Declaration of Human Rights (1948) available at https://www.ohchr.org/EN/UDHR/Documents/UDHR_Translations/eng.pdf (accessed: 6 December 2018).

United States (2009) 'Observations by the United States of America on the relationship between climate change and human rights' available at www.ohchr.org/Documents/Issues/ClimateChange/Submissions/USA.pdf (accessed: 14 March 2017).

Waldron, Jeremy (2003) 'Indigeneity? First peoples and last occupancy', *New Zealand Journal of Public and International Law*, 1(1): 55–82.

Wetherell, Margaret & Potter, Jonathan (1988) 'Discourse analysis and the identification of interpretive repertoires', in Antaki, C. (ed.) *Analysing Everyday Explanation*, London: Sage: 168–183.

Wilson, Richard Ashby (2006) 'Afterword to "Anthropology and human rights in a new key": The social life of rights', *American Anthropology*, 108(1): 77–83.

World Bank (2011) 'Human rights and climate change: A review of the international legal dimensions' available at http://siteresources.worldbank.org/INTLAWJUSTICE/Resources/HumanRightsAndClimateChange.pdf (accessed: 25 July 2016).

World Bank (2015) 'Environmental and social framework: Setting environmental and social standards for investment project financing' (1 July) available at https://iphrdefenders.net/world-bank-environmental-social-framework-setting-standards-investment-project-financing/ (accessed: 9 July 2018).

World Bank Group (2017) 'Water, well-being and the prosperity of future generations' available at https://openknowledge.worldbank.org/handle/10986/26203 (accessed: 2 August 2017).

World Commission on Environment and Development (1987) *Our Common Future*, Oxford: Oxford University Press/United Nations.

Chapter 4

Balancing generational sovereignty with a future ethics

Introduction

This chapter examines the growing tendency amongst concerned citizens and environmental campaigners around the world today to pursue climate justice through the courts. One common viewpoint shared by all of these legal campaigners is the notion that government agencies have become distant and too far removed from the ecological, social and human rights concerns that preoccupy contemporary publics. In many ways, these campaigns grow from deeper sentiments of enduring disappointment with state democratic life and the diminishing capacities of government administration to secure ecological, political and economic advantages for all peoples. Youth, in particular, ask how can 'we the people' preserve a sense of our democratic capacity when everywhere there is evidence of its demise? Continuously, publics are reminded of the contradictions that prevail between justice principles and the reality of everyday living. Critical reflection on the distance between them always brings justice back to its actual basis, the social agent (Touraine, 1988) and the desire to ensure that justice remains the work of the people, not some purely grand narrative (Cornell, 1990). Insofar as this orientation towards justice offers a defense of the necessity, locally, nationally and internationally, of campaigns for democratic reform, accountability and inclusion, it also shows that even today, in this era of Trump and Viktor Orbán, justice continues to be contested, fought over and defended by newer generations.

Before discussing specific legal campaigns brought against states for poor performance of duties to protect the environment for present and future posterity, the following analysis outlines some wider developments in international political and legal discourse on climate change that have had major impact on how courts respond to claims linking human rights violations to environmental degradation.

In the years since representatives of Small Island Developing States adopted the Malé Declaration on the Human Dimensions of Global Climate Change (November 2017), the first intergovernmental statement recognizing the implications of climate change 'for the full enjoyment of rights', the relationship between human rights and climate change has become a more focused concern of various international bodies (e.g., the Human Rights Council (Resolution 35/20, 2017;

Resolution 31/52, 2016) and the Office of the High Commissioner for Human Rights, 2016). However, it was not until the Paris Agreement, signed in December 2015, that a truly effective remedy for the violation of protected rights could be envisaged. Each sovereign signatory to this agreement was now expected to take strategic action to ensure that the procedural and substantive obligations bestowed on it under human rights law (OHCHR, 2015) would be realized. In the Preamble of the Paris Agreement, states are reminded of their duty to:

> [r]espect, promote and consider their respective obligations on human rights, the right to health, the rights of indigenous peoples, local communities, migrants, children, persons with disabilities and people in vulnerable situations and the right to development, as well as gender equality, empowerment of women and intergenerational equity.
>
> (Paris Agreement, FCCC/CP/2015/L.9, 20 [December 12, 2015])

This requirement represents an important step towards a greater legal clarification of the obligations and responsibilities of states to address the human rights challenges of climate change. With such clarification comes a stronger legal basis for challenging any state that does not fulfill these obligations. One might question the notion of holding individual states accountable for changes that in every way are global, created largely as they are by the activities of transnational corporations and consumers worldwide. Sources of climate harm today have become more diffuse, with developing economies such as that of Brazil now officially the fifth largest polluter in the world and India the eighth (World Economic Forum, 2016). However globalized sources of climate destruction have become, this fact has not deterred concerned citizens and environmental groups from reminding states of their obligations to take immediate action to reduce CO_2 emissions levels. As a global challenge necessitating a 'global response' (Article 2, The Paris Agreement, 2015), climate change bears directly on the responsibilities of all sovereign states to preserve the essential resource base upon which their peoples depend for their survival and future flourishing. However, it is not only international agreements on climate change mitigation that compel more immediate action on the part of states. Campaigners also remind states of established legal doctrines, obliging states to hold common resources in trust for present and future generations (the preservation of lands in their natural state, the protection of scenic and wildlife habitats, as well as waterways for recreational uses and fisheries for essential food stocks, etc.). This doctrine prohibits the sale of such resources for development purposes or their use in ways that jeopardize common interests (e.g., see Indian Supreme Court [1996] *M.C. Mehta v. Kamal Nath*, (1997) 1 S.C.C. 388; in Canada, *Labrador Inuit Ass'n v. Nfld. [Minister of Env't and Labour]*, [1997] 155 Nfld. & P.E.I.R. 93 (Can.); *U.S. Supreme Court, Illinois Central Railroad v. Illinois*, 146 US 387 [1892]; see also Blumm & Guthrie, 2012).

The fact that many states regularly do lease or sell what were once common resource assets in no small way contributes to their record of extraordinary failure in environmental protection. Frustrated by this state of affairs, concerned citizens turn to law in an attempt to halt what is largely seen as a government-orchestrated demise of their ecological heritage. Government agencies are accused of misusing the latitude granted to them in their role as regulatory agents. Public trust in state agencies rests on one core assumption – that these agencies will exercise a neces-sary degree of precaution in the protection of precious public assets. For increas-ing numbers, state bodies do not practice sufficient precaution but instead have fallen captive to the charms of the industries they regulate. Campaigners highlight how ongoing natural resource depletion is not an issue that states can choose to ignore or downplay, especially when they have played a key role in bringing about such destruction. Instead, climate change is seen as a reflection of the failures of states to fulfill basic sovereign duties, especially the duty to protect the interests of their citizens. Indeed, it is this diagnosis that prompts growing numbers to pursue corrective action via other means, most noticeably the courts. These actors challenge governments in their role as environmental protectors, claiming that their activities violate not only the details of international agreements (including procedural rights to public participation [e.g., UNFCCC, Article 4(1)(i); Aarhus Convention, Article 6] and the right to health [e.g., see International Covenant on Economic, Social and Cultural Rights (1966), Article 12]) but also constitutional law and the public trust doctrine.

Critics point to the fact that, as trustees, states cannot sit idly by and let essential resource asset depletions continue. Rather, all sovereign states around the world are connected to one another and to the planet as a whole as cotrustees of shared ecosystems. Beyond duties owed to their own respective citizens, states are bound by legal duties to global others not to waste common assets, such as migratory fish stocks, the waters of transboundary rivers, biodiversity, and so on. Fiduciary obligations to protect the global atmosphere, similarly, necessitate states to take measures to reduce carbon emissions. Indeed, a whole body of international and domestic law supports an institutionally embedded trust framework that strategi-cally positions legal courts as enforcers of fiduciary duties, thereby ensuring that state bodies manage common assets sustainably. The powers of legal courts in this regard has been further reinforced by the terms of the Paris Agreement (2015) and the call for a more vigorous judicial review of states' environmental performance (e.g., Woods, 2014b: 538). A public trust doctrine in a post–Paris Agreement era emphasizes states' position within a legally monitored sovereign framework as natural resources trustees (Woods, 2014a), holding them accountable for any fail-ures in this function. Given its explicit reference to 'intergenerational equity', the Paris Agreement also interprets this doctrine as applicable across generations.

'The Earth belongs always to living generations': how Jefferson got it wrong

In a letter to James Madison in 1789, Thomas Jefferson ([1789] 1905) laid out an argument on the relationship between generations that continues to resonate with many today. Here Jefferson explains how the freedom and sovereignty of living generations must not be compromised by debts accumulated by past generations. Equally for Jefferson, it is important we do not become wedded to the notion that the Earth and its resources belong to generations not yet born. The only thing owed to future generations, according to Jefferson, is a wide and clean space for action. With 'the earth belong[ing] always to living generations, they may manage it then and what proceeds from it, as they please, during their usufruct'. The function of the state in this instance is to protect the interests of the living because its legitimacy and authority depend upon their consent. So understood, the sovereignty of a people is a living presence, a spirit of revolution that becomes part of a history still in the making. To maintain the vibrancy of each generation's sovereignty, Jefferson suggests that legislative measures be limited to the time frame of each generation's governing period. Once that governing period ends, legal rules should once again be opened up to amendment by the next governing generation:

> [T]he earth belongs to each of these generations during its course, fully, and in their own right. The 2d. generation receives it clear of the debts and incumbrances of the 1st., the 3d. of the 2d. and so on. For if the 1st. could charge it with a debt, then the earth would belong to the dead and not the living generation. Then no generation can contract debts greater than may be paid during the course of its own existence.
>
> (Thomas Jefferson, Letter to James Madison on September 6, 1789, Correspondence [1789–1792] (1905))

At the core of Jefferson's thinking is the notion of maximizing the liberties of each generation to shape the world in a way they desire. However, this approach does not consider the consequences that arise across time of each generation's efforts to shape the world as they wish and according to their own interests, especially when that means depleting reserves of natural resources to satisfy carbon-intensive lifestyles so that insufficient amounts of essentials are left for those that come after. There is no space in Jefferson's thinking for the notion of limited natural resource opportunities or ensuring that 'enough and as good' be left for newer generations (Locke, [1689] 1988). We are only beginning to understand the dire consequences of thinking purely of the resources needs of the present, as well as the importance of ensuring that the sovereignty of each generation is determined in a more reflective manner, that is, in ways that extend consideration to how our actions and choices will shape the circumstances of newer generations. It is not only a matter of protecting the sovereignty of each generation (as Jefferson suggests) but, even more fundamentally, protecting the capacity of each to exercise sovereignty by ensuring that necessary resources are available for generations to be self-determining. As climate conditions continue to deteriorate and global reserves of essential resources, such as freshwater and arable lands become less plentiful, prospects for future generations exercising sovereignty in an unproblematic way become less predictable. While Jefferson's intentions clearly were to avoid scenarios where the freedoms of each new generation are unfairly restricted by long established rules, such as an inherited constitution or accumulated debt, the reasonableness of prioritizing living generations over future ones is not self-evident. In his reply to Jefferson, James Madison pointed to the fact that generations do not supersede one another in a neat manner but rather overlap, thereby making a clear differentiation between their interests less straightforward. There is also the fact that the world and its resources are not 'owned' by generations 'taking possession' of them, as Jefferson implies but, rather, are received as a gift. The '*improvements* made by the dead form a debt against the living who take the benefit of them' (Peterson, 1976). Thus as beneficiaries of the preservation efforts of past generations, living generations are obliged to make further 'improvements' rather than impose disimprovements (through practices of environmental degradation) on those that come after. Such obligations condition our freedom to use the resources of this world as we wish, as well as limit our freedom (somewhat) to assert only our own interests.

As natural resource destruction intensifies, the contemporary relevance of this historical debate between Jefferson and Madison is striking. Debates on the question of generational sovereignty rage once again, but this time the stakes are much higher and the strategies used by campaigners to control the activities of living agents of harm more far-reaching. The realization is that the short-term gains to be had from burning remaining fossil fuels reserves, for instance, will seriously compromise the ecological, social and economic circumstances of future generations (Gosseries, 2008). The critical nature of contemporary ecological problems leads growing numbers to turn to law in the hope that the legal system can impose sanctions on the conduct of high emitting agents and to take governments to task for not controlling rates of resource depletion:

> [T]he cost to human life [of ongoing environmental degradation] necessitates a need for the courts to evaluate the constitutional parameters of the action or inaction taken by the government. . . . [G]lobal warming will eventually hurt us all but it will hurt our children and grandchildren the most.
> (Judge Thomas Coffin, United States Federal District Court in Eugene, Oregon, 2015)

Legal actors such as the Law Response Team, based in London, The Hague, and New York, operate as independent advisors to those wishing to pursue charges of ecocide against large corporations or challenge governments legally in their decisions regarding pollution control. The understanding is that 'extensive damage to, destruction of, or loss of ecosystem(s) in a given territory, whether by human agency or by other causes, to such an extent that peaceful enjoyment by the inhabitants of that territory has been, or will be severely diminished' (Eradicating Ecocide, 2015) is a crime against present and future humanity (see also World Future Council, 2013). In October 2016, an international public tribunal was staged in Le Hague, the Netherlands, holding Monsanto accountable for acts of genocide against the ecosystems upon which human life depends (Chair of this tribunal, Judge Tulkens). Over a two-day period, a panel of experts, including consultants to the International Criminal Court, Dior Fall Sow and Jorge Fernandez Souza, heard the testimonies of multiple victims (Barroux, 2016). On April 18, 2017, the chair of the tribunal and former vice-president of the European Court of Human Rights, Judge Tulkens (Wijdekop, 2016), publicly delivered the tribunal's final judgment. The purpose of this tribunal, according to Judge Tulkens, was to 'give a voice to witnesses' and gather empirical evidence of acts of ecocide committed by those who are liable for prosecution under international law'. For its part, Monsanto refused to participate in the tribunal, describing it as 'a mock trial' with no 'legitimate connection' to the international legal order. In an open public letter published in October 2016, Monsanto explained how it remains committed to protecting human rights and continues to follow the UN's Guiding Principles for Business and Human Rights.

For those campaigning for a legal recognition of the crime of ecocide, activities that destroy the environment infringe on the human right to life and security and, for such reasons, ought to be considered on par with other international crimes, that is, crimes that pose a serious threat to 'the peace, security and wellbeing of the world' (preamble to the Rome Statute of the International Criminal Court). International justice, these actors argue, must play a key role in providing the basic conditions for global sustainable development and insisting upon the pre-eminence of the rule of law when addressing sources of pollution threat (e.g., state parties and major corporate actors). If a legislative change was introduced, offending parties would be rendered legally liable for ecocide crimes in accordance with the three principal modes of liability: aiding and abetting, joint criminal enterprise and superior responsibility. Proposals to extend legal interpretations of aggression to cover large-scale environmental destruction have been circulating for more than four decades now. In 1972, Olof Palme, then prime minister of Sweden, in his opening speech at the Stockholm Conference for the Human Environment, referred to the Vietnam War as an 'ecocide'. Indeed, it was this international conference that highlighted the transnational relevance of growing environmental problems. A Working Group on Crimes against the Environment was formed at the conference, and a draft Ecocide Convention was submitted to the United Nations in 1973. The initial plan was to include the crime of ecocide in drafts of the Rome Statute (1985–1996), but this entry was subsequently removed. The understanding was that ecological crimes could be added in the future, if conditions necessitate it, in line with the continuous institutional development of a notion of individual criminal accountability. Historically, the latter has led to an expansion of categories of serious crimes, from piracy and war crimes, followed by crimes against humanity, aggression and genocide in the postwar era, and later extending to the crime of apartheid and torture.

As global climate conditions continue to decline, it is perhaps not surprising to see international support for such proposals growing once again. In September 2016, the International Criminal Court in The Hague decided to include 'destruction of the environment' and 'illegal exploitations of natural resources' for the first time in its field of criminal investigations, possibly in recognition of the increasing number of legal agencies pursuing environmental cases. It also decided to broaden the range of victims to include 'indirect victims'. Even though the possibility of a formal recognition of environmental destruction as a global crime is still in its early stages of development, nonetheless, new initiatives do signal a change beginning to occur in legal institutional thinking. Also, some progress has been made through legal courts in the prosecution of states who fail to meet agreed CO_2 emissions targets. An outline of some of these cases is now presented.

Pursuing climate justice through the courts: *the people v. the state*

In Belgium the nonprofit agency Klimaatzaak initiated a climate change lawsuit against the governments of Flanders, Wallonia, Brussels and the Federal Belgium government in April 2015, demanding that they reduce greenhouse gas emissions by 40 percent before 2020 and steer Belgium's transition to a low-carbon future. The attorney acting in this case, Roger Cox, also acted as legal counsel for Urgenda, an environmental organization that brought a liability suit against the Dutch state. In June 2015, the District Court of The Hague ruled that the Dutch government's plan to lower its emissions reduction target for 2020 by 14–17 percent compared to 1990 levels was 'unlawful' and contrary to international agreements on climate change mitigation. The state was ordered to reduce greenhouse gas emissions by 25 percent within the next five years.[1] In delivering its judgment, the Court cited the European Convention on Human Rights, the Dutch constitution, and the principle of 'no harm'. The following December, the Netherlands, along with 174 other states signed the UN Climate Treaty in Paris and once again expressively accepted its responsibilities to ensure that national emission levels be reduced to prevent further ecological regression. The *Urgenda v. The Kingdom of the Netherlands* case, along with the Klimaatzaak lawsuit in Belgium, offers important insights on how states in the future might be compelled to recognize legal responsibilities to lower emissions levels and acknowledge the effects of their activities on vulnerable communities, including future generations. Already, these cases have inspired similar legal cases elsewhere. In September 2015, Ashgar Leghari, a farmer from Pakistan, using public interest litigation (PIL), successfully sued the government of Pakistan for delaying the implementation of its National Climate Change Policy (2012 and, by doing so, violating the constitutional rights of citizens. Acting Judge Syed Mansoor Ali Shah of the Lahore High Court Green Bench ordered the establishment of a national Climate Change Commission with a clear remit to ensure effective implementation of the 2013 policy plan.

In its first order, issued September 4, 2015, the Court noted:

> Climate change is a defining challenge of our time and has led to dramatic alternations in our planet's climate system. For Pakistan, these climate

variations have primarily resulted in heavy floods and droughts, raising serious concerns regarding water and food security. On a legal and constitutional plane, this is clarion call for the protection of fundamental rights of the citizens of Pakistan. In particular, the vulnerable and weak segments of the society who are unable to approach this Court.

(Lahore High Court Green Bench, 2015)

In paragraph 6, the Green Bench invoked the right to life and the right to dignity embedded in the constitution of Pakistan, as well as the precautionary principle as necessary guides to achieving greater 'climate change justice' (para. 7). Concluding that federal and provincial officials have done little to date to implement adaptation measures to cope with changing climatic patterns, the Green Bench directed responsible ministries and departments to appoint a focal person that would appear before the Green Bench on issues relating to climate change and prepare a list of adaption measures that will be implemented before the end of 2015. The Court also established a Climate Change Commission to help it monitor progress in this regard and to ensure compliance with its guidelines (see para. 8). In a second order issued September 14, 2015, the Green Bench stressed the importance of implementing the recommendations of Pakistan's National Climate Change Policy 'to ensure climate change is mainstreamed in economically and socially vulnerable sectors' (para. 9). Present in Court on this occasion were representatives from the government's Cabinet Division, the Ministry of Finance, Revenue and Planning, the Ministry of Foreign Affairs, the Ministry of Inter-Provincial Coordination, the Ministry of Climate Change, the Ministry of Water and Power, and the National Disaster Management Authority.

In the period since Leghari brought this case against the state of Pakistan, Urgenda against the Dutch state or Klimaatzaak against the Belgian state, concerned citizens in other countries, including France and Norway, have begun to prepare similar cases, demanding legal accountability through the courts for rising CO_2 emissions levels, as well as an immediate implementation of measures to reduce rates of pollution. In the process, citizen agency is extended, and wrongful acts of environmental destruction are linked in a more tangible way to responsible agents. Traceable links are forged between rights recipients who have suffered an identifiable or concrete injury, on the one hand, and government agencies appointed by the state to fulfill duties of care on the other. As scientific knowledge of harm grows more precise (e.g., The Lancet Commission's reports on pollution and health [2017]; the European Environment Agency [2016] on correlations between air pollution and premature deaths) and the effects of climate change more immediate, demands for accountability and redress increase in equal measure (see John H. Knox, 2009; UNEP Report, 2015: 13).

Legal cases brought against state departments for poor performance of environmental duties suggest a growing level of public frustration with their lack of commitment to climate change mitigation. They also suggest that other means of problem resolution are failing, leaving citizens to pursue legal means as a last resort. There is, however, one development in this regard that is particularly noteworthy, and that is the increasing number of legal cases filed by youth against their states. Amongst other legal obligations, youth campaigners draw attention to violations of Children's 'inalienable rights to a stable climate system' (Julian Olson, Our Children's Trust, 2016), their rights to survival and development (Just Planet, 2016: 17) and the importance of corrective actions to safeguard their future. Evidence of 'particularized injuries' is linked to specific sources of harm (i.e., time-determinate harms) over which they are denied some degree of control. Emphasis is also placed on shared responsibilities for ongoing unjust outcomes in the long-term distribution of climate change effects (forward-looking interpretations of justice). For youth campaigners, what is sought is something akin to Young's (2011) 'social connection model of responsibility' for climate injustice, where remedying the effects of adverse climate conditions is understood as requiring immediate action that must continue thereafter, implicating multiple agents across space and time, as beneficiaries and duty bearers in the search for effective long-term solutions.

Legal challenges exploring the 'plural temporalities' (Young, 2011: 108) of environmental responsibility (encompassing present and future generations) include the civil case brought by 44 minors in the Philippines in 1990. Represented by their parents and leading lawyer, Antonio Oposa, (n. 90–777) at the Regional Trial Court of Makati, Metro Manila, these youths filed a class action against the then secretary of the Department of Environment and Natural Resources (DENR) demanding the immediate cessation of all existing timber license agreements in the country, as well as approval of new ones. The Regional Trial Court subsequently rejected the plaintiffs' case on the grounds that they lacked juridical personality. The nature of their grievance (the harming of younger generations and those not yet born) was thought to be of a political nature, that is, unsuitable for the Regional Trial Court to decide upon. On appeal, the case passed to the Supreme Court (*Minors Oposa v. Secretary of the Department of Environment and Natural Resources [DENR]*). In July 1993, this court ruled in favor of the plaintiffs, overturning the decision of the Manila Regional Court and ordering the Department of Environment and Natural Resources to cancel all existing timber license agreements in the country on the grounds that they encourage practices that violate constitutional rights to a healthy environment by destroying resources 'held in trust for the benefit of plaintiff minors and succeeding generations' (Minors Oposa, supra note 65, 180). The Department was also ordered to cease and desist from receiving, accepting, processing, or approving new timber license agreements' on the grounds that such agreements allowed for the imposition of 'great damage and irreparable injury' on plaintiff minors 'and their successors who may never see, use, benefit from, and enjoy this rare and unique natural resource treasure'. The Supreme Court recognized the Plaintiffs 'clear and constitutional right to a balanced and healthful ecology', as well as their entitlement to protection by the state in its capacity as the *parens patriae* (Republic of the Philippines Supreme Court, Manila, G.R. No. 101083 July 30, 1993, *Minors Oposa*, supra note 65, 180).

Of crucial importance in this case was the Supreme Court's acknowledgment of the failure of the state to fulfill its duties to protect the community from long-term ecological damage. Second, the right of minors to file a class suit on behalf of themselves and generations not yet born (Opinion of Associate Justice Hilario G. Davide Jr., *Minors Oposa*, supra note 65, 185). The opinion of the court was that the plaintiffs had put forward compelling evidence documenting how, 25 years earlier, the Philippines had 16 million hectares of rainforests (53 percent of its landmass), but by the fall of the old regime in 1986, the rainforests accounted for just 1.2 million hectares (4 percent) of the country's landmass, a situation considered highly dangerous for tropical environmental conditions in the years ahead. Indeed, this evidence was crucial to the plaintiff's argument and clarified precisely how the policy of the state was negatively impacting the life chances of future generations. In its ruling, the Supreme Court quoted the Constituent Commission set up after the revolution of February 1986 and the formation of a new Republic of the Philippines in its acknowledgment of the importance of 'judicious management and conservation of the country's forests' without which 'environmental balance would be irreversibly disrupted'. At the time, the Supreme Court's ruling was exemplary in its recognition of the possibility of a class action, where legal rights and responsibilities (e.g., the 'right to a balanced and healthy ecology', Section 16 of Article II of the 1987 Constitution) are legitimately extended to future generations. The nonidentity of presently unborn future peoples (who did not exist at the time of this legal hearing) did not prove a problem for the court. Neither did the lack of contemporaneity of the harm doer and the majority of victims. Instead, emphasis was placed on the relevance of legal protections across time. Not only did this court victory prove an important reminder to states of the significance of intergenerational components of environmental legislation, it has also been a major inspiration to other youth coalitions across the world seeking legal justice.

On August 12, 2015, 21 youths (aged 8 to 19), together with climate scientist James E. Hansen, filed a complaint against the federal government in the U.S. District Court in Oregon, alleging that it had violated 'the fundamental right of citizens to be free from government actions that harm life, liberty, and property' chiefly by 'approving and promoting fossil fuel development, including [the] exploration, extraction, production, transportation, importation, exportation, and combustion' of fossil fuels (Complaint for Declaratory and Injunctive Relief at 85, *Kelsey Cascadia Rose Juliana, Xiuhtezcatl Tonatiuh M. Et Al. v. United States, Barack Obama et al.*, No. 6:15-cv-01517-TC, August 12, 2015).

According to the plaintiffs in this case, the federal government had violated substantive due process rights by allowing atmospheric CO_2 emissions to reach levels that will endanger their lives, liberty and property, as well as that of future generations. The plaintiffs also claimed the right to equal protection against dangerous levels of atmospheric pollution. Such rights, they argued, have been breached by the decision of the federal government to approve projects such as the Jordan Cove Liquefied Natural Gas Project in Oregon. Eighteen-year-old plaintiff and Oregon resident Alex Loznak described his injury with respect to the Jordan Cove project:

> By 2020, [the project] will be the largest emitter of greenhouse gases in the whole state of Oregon. Science tells us we must sharply cut back on CO_2 emissions, but my Federal Government has given the green light to massive LNG exports from this terminal. If constructed, the terminal would process one billion cubic feet of natural gas per day, locking us into dependence on fossil fuels at a time when we should be transitioning toward a renewable energy economy. My family has owned a farm near the proposed pipeline route for almost 150 years, and I'm worried about the impact that increased drought and wildfire will have on the farm unless we act now on climate change.
>
> (Our Children's Trust, 2017)

In total, 96 pages of the plaintiffs' complaint outlines in detail how their rights to life, liberty, property and equal protection are adversely affected by new fossil fuel projects. The text establishes that, as young people, the plaintiffs are 'especially vulnerable to the dangerous situation that the defendants [government] have substantially caused'. Julia Olsen, lead counsel on this case and public interest attorney for Our Children's Trust, asked the court if 'our government has a constitutional responsibility to leave a viable climate system for future generations?' Similar to arguments presented during the *Minors Oposa v. Secretary of the Department of the Environment and Natural Resources* Supreme Court case in the Philippines (1993), government was continuously reminded of its duties to protect citizens, young and old, present and future. The federal government and fossil fuel industry representatives made a motion to dismiss the plaintiffs' complaint, both of which were subsequently rejected by Federal Court Judge Thomas Coffin in April 2016 when Coffin reminded government of its obligations to comply with 'the public trust doctrine' and fulfill its duties to act as a trustee of the natural resources upon which the community depends (Conca, 2016; Blumm & Wood, 2013). In September 2016, U.S. Magistrate Judge Aiken reviewed and upheld the decision of Judge Coffin and issued a ruling to that effect in November 2016.

To satisfy the court's requirements, the plaintiffs were asked to document how they have been injured in a 'personal and individual way' by projects such as the Jordan Cove Liquefield Natural Gas Project. They were also asked to demonstrate how in seeking relief, they will 'directly and tangibly' benefit in a manner that is distinct from 'the public at large'. The plaintiffs have already submitted details on how the livelihoods and living environments of these 21 youths are injured in a 'particularized' manner. Eighteen-year-old Jacob Lebel, for instance, offers an account of how he 'is harmed and will continue to be harmed' by the defendants' actions since environmental degradation affects the farm on which he works and his family's ability to secure a livelihood. Similarly, Zealand Bell, 11, describes how increased drought, heat waves and warmer temperatures threaten his 'enjoyment of outdoor activities' and have led to a loss of income for his family, as his mother's seasonal job at a ski resort was not available in 2014 due to lack of snow. Sahara Valentine, also 11, explains how she has experienced several asthma attacks from the increased frequency of forest fires in Oregon brought on by the drier, hotter weather conditions. In this case, the plaintiffs' claims of harm were taken seriously by the court. In other cases, however, the need to establish traceable links between sources of pollution and their effects have proven too great a task. For instance, in the case *Washington Environmental Council v. Bellon* (2013), the U.S. Court of Appeals for the Ninth Circuit held that those who had brought the case against the state of Washington (Washington Environmental Council, Sierra Club, Washington State Chapter) had failed to demonstrate how emissions from the five oil refineries under review had imposed injuries upon them directly or had made a 'contribution to global GHG levels'. The need to provide evidence of harm has, however, not deterred other youth campaigners from filing complaints.

Six members of the Boulder-based youth group Earth Guardians (Xiuhtezcatl Martinez, Itzcuahtli Roske-Martinez, Sonora Brinkley, Aerielle Deering, Trinity Carter and Emma Bray) filed a petition in Boulder, Colorado, in November 2013, asking the Colorado Oil and Gas Conservation Commission to introduce a new rule to stop the practice of fracking until such time as it can be proven that this and other forms of gas and oil extraction do not impose adverse harms on 'the health and safety of Colorado's residents and the integrity of Colorado's atmospheric resource and climate system, water, soil, wildlife, other biological resources' (Petition to the Colorado Oil and Gas Conservation Commission and Colorado Department of Natural Resources, November 15, 2013: 3). According to the evidence presented by the plaintiffs, Colorado already emits higher quantities of greenhouse gases, including carbon dioxide, methane and aerosols, than 174 countries around the world and that the oil and gas sector in Colorado is a major contributor to this problem. The plaintiffs reminded the court of the duties of the state to protect natural resource assets and, in doing so, protect their right to a 'livable future':

> A livable future includes the opportunity to drink clean water and abate thirst, to grow food that will abate hunger, to be free from imminent property damage caused by extreme weather events, and to enjoy the abundant and rich biodiversity on this small planet.
>
> (Ibid.: 94)

Petitioners called on the Commission to fulfill its legal responsibilities to protect 'the basic human right to clean water, clean air, and a healthy future for all young people and generations to come' (Earth Guardians and Our Children's Trust, June, 2014). On May 29, 2014; the Colorado Oil and Gas Conservation Commission issued a written statement denying the antifracking petition of these young campaigners, concluding that 'other Commission priorities . . . must take precedence' over the youth petitioners' demands (ibid.: 1). On February 19, 2016, the Denver District Court affirmed Colorado Oil and Gas Conservation Commission's order to deny the fracking petition brought by these young plaintiffs. However, on July 3, 2014, the petitioners appealed the decision of the Commission and that of the District Court pursuant to the State Administrative Procedure Act.

On March 23, 2017, Judge Fox of the Colorado Court of Appeals reversed the Colorado Oil and Gas Conservation Commission's order denying the plaintiffs' petition against fracking and the district court order upholding the denial. Judge Fox remanded the case to the district court and noted that the Commission had misinterpreted Colorado law:

> [T]he Commission erred in interpreting [the Oil and Gas Conservation Act] as requiring a balance between development and public health, safety, and welfare. . . . [T]he clear language of the Act . . . mandates that the development of oil and gas in Colorado be regulated subject to the protection of public health, safety, and welfare, including protection of the environment and wildlife resources.
>
> (Colorado Court of Appeals, 2017)

The case was remanded to the district court to return it to the Commission for further proceedings consistent with this opinion. According to the Plaintiff's counsel, Julia Olson, executive director of Our Children's Trust, the decision of Judge Fox and the Court of Appeals highlighted the Commission's wrongful prioritization of further gas and oil development in the State of Colorado over the health and safety concerns of citizens. Campaigners interpreted this move by Judge Fox as a much needed vote of confidence in youth and their moves to establish a tighter regulatory environment.

> The Commission can no longer decide to prioritize oil and gas development over the health and safety of Coloradans. This is an enormous victory for these youth. We look forward to helping the youth of Colorado go back before the Commission on remand.
>
> (Our Children's Trust, March 2017)

On November 19, 2015 Judge Hollis R. Hill issued a groundbreaking ruling in the case of eight youth petitioners (*Zoe and Stella Foster, et al. v. Washington Department of Ecology*, 2015) who had requested the Washington State Department of Ecology to introduce new carbon emissions legislation protecting the atmosphere for generations to come. Judge Hill reminded the Washington State Department of Ecology of the public trust doctrine that mandates the state to act through its designated agency to protect the natural environment from harm. As Judge Hill noted, the state 'has a constitutional obligation to protect the public's interest in natural resources held in trust for the common benefit of the people' (Supreme Court of the State of Washington for King County, Ruling of Judge Hollis R. Hill,

2015). The Court upheld the young petitioners' claim that 'the scientific evidence is clear that the current rates of reduction' mandated by Washington Law 'cannot ensure the survival of an environment in which [youth] can grow to adulthood safely'. Judge Hill concluded that future generations' 'very survival depended on the will of their elders to act now, decisively and unequivocally, to stem the tide of global warming . . . before doing so becomes first too costly and then too late'. More specifically, the state has a 'mandatory duty' to 'protect and enhance air quality for current and future generations' (Judge Hill, *Zoe and Stella Foster, et al. v. Washington Department of Ecology*, 2015). Following Judge Hill's ruling, the Department of Ecology was instructed to impose a more stringent limit on carbon emissions in the State of Washington. In February 2016, the Department withdrew its new proposed rule on emissions standards, prompting the youth petitioners to ask the court to step in again and impose a court order on the State of Washington's Department of Ecology to reduce emissions limits (April 2016). Superior Court Judge Hollis Hill ordered the Department of Ecology to promulgate an emissions reduction rule by the end of the year and make recommendations to the state legislature on science-based greenhouse gas reductions in its 2017 legislative session.

On June 1, 2016, the Department of Ecology released its revised Clean Air Rule. According to the young petitioners in this case, the new rule was based on outdated emissions figures and required an annual reduction of 1.7 percent in emissions levels, which they considered inadequate. On July 14, 2016, the Department of Ecology held a public hearing on the proposed Clean Air Rule, as ordered by the court. Oral and written comments were invited from the public, and on July 25, 2016, representatives of the youth petitioners in this case, Julia Olson (Chief Legal Counsel at Our Children's Trust) and Andrea Rodgers (Western Environmental Law Center [WELC]) submitted comments requesting the Department of Ecology to bring their proposed Clean Air Rule into compliance with the law. A blueprint was also prepared outlining how the State of Washington could be put on a more sustainable path of development. On November 22, 2016, youth petitioners returned to court for a hearing. Governor Jay Inslee of the Department of Ecology presented a case in defense of his department, claiming it was not in contempt of the court's prior order to protect the constitutional rights of youth. On December 19, 2016, Judge Hollis Hill ruled that the youth petitioners could again move forward with 'constitutional climate rights claims', adding the State of Washington and Governor Jay Inslee as defendants in the case (Superior Court of the State of Washington, Judge Hollis R. Hill, December 19, 2016). Although this case is still ongoing, progress to date suggests a greater willingness on the part of the court to hold state offices accountable for rising emissions levels and the need to protect citizens' rights.

In January 2018, 25 plaintiffs, aged 7 to 26, together with human rights group Dejusticia, brought the first Latin American lawsuit defending the rights of youth and future generations before the Superior Tribunal of Bogotá in Colombia. The primary grievance was the ongoing deforestation of the Colombian Amazon (rates of deforestation in the Amazon region increased by 44 percent between 2015 and 2016 alone). According to the plaintiffs in this case, government's failure to control rates of deforestation poses a serious risk to the safety, health and sustainable living conditions of future generations, as well as a serious violation of agreements under the Paris Agreement (2015) to reduce rates of deforestation. The April 2018, the highest court in Bogotá, ruled in favor of the plaintiffs and instructed the government to take immediate action to protect Colombia's Amazon rainforests and reduce rates of deforestation. The acting Supreme Court Judge described how:

> [i]t is clear, despite numerous international commitments, regulations . . . that the Colombian state has not efficiently addressed the problem of deforestation in the Amazon.
>
> (Moloney, 2018)

At both the local and national level, government authorities were ordered to prepare an action plan on how it will combat deforestation and better protect 'rights for our children or for future generations'.

In Europe also, the number of court actions taken by youth and their representatives is increasing, especially since the signing of the Paris Agreement in December 2015, when states pledged to take necessary actions to reduce their CO_2 emissions levels and satisfy human rights obligations to protect especially the most vulnerable, such as children. In Norway, in October 2016, a group of young citizens (members of the environmental group Nature and Youth), together with Greenpeace Norway and Grandparents Climate Campaign, filed a constitutional climate lawsuit against the Norwegian government for allowing oil companies to drill in search of new reserves of oil in the Arctic Barents Sea and, in doing so, burdening present and future generations with further environmental destruction.[2] Inspired by similar youth-led legal actions elsewhere, these campaigners claim that the Norwegian government has violated citizens and future generations' constitutional right to a healthy environment. In particular, their grievances are focused on the decision of the Norwegian government to offer new production licenses to oil and gas exploration companies in June 2016 (including Statoil, Chevron and Aker BP). For the first time in more than 20 years, Norway opened new acreage in northerly and easterly areas of the Barents Sea for oil exploration just six months after signing the Paris Agreement (Plaintiffs' Statement, October 18, 2016). The plaintiffs in this case questioned the constitutional (Norwegian Constitution, 1814) and procedural validity of the government's licensing decision, as well as the adequacy of its commitment to fulfill duties of care to today's youth and generations to come (Nature and Youth, 2017). 'Norway must contribute to a reduction in global greenhouse gas emissions (Greenpeace International, 2017). It is impossible to reconcile the petroleum production permitted by the licensing decision with the reduction in emissions that Norway must contribute in order to avoid devastating and irreversible climate change' (ibid.: 1). According to the plaintiffs, the Norwegian government, in issuing new exploration licenses, is guilty of violating Article 112 of the Norwegian constitution, which reads as follows:

> Natural resources shall be managed on the basis of comprehensive long-term considerations which will safeguard this right for future generations as well. The authorities of the state shall take measures for the implementation of these principles.

On the first day of court proceedings, Greenpeace placed a five-ton block of ice outside the court premises in Oslo with Section 112 of the Norwegian constitution carved into it to remind government that it is placing Norway's constitutional commitments to intergenerational justice in jeopardy. Also noted are commitments to the European Convention on Human Rights (Our Children's Trust, 2016). Amongst those testifying in court were Pacific Islanders acting as representatives of peoples most affected by climate change. The demand is for immediate corrective action to be taken, as well as a justification from the state as to how this situation arose in the first instance. Ingrid Skjoldvær, from the Nature and Youth group, added: 'We will argue in court that the Norwegian government has an obligation to keep its climate promises and will invoke the people's right to a healthy environment for ours and future generations. This is the People v's Arctic oil!' (quoted in Neslen, 2016). On January 4, 2018, the Court ruled that the Norwegian government was not guilty of breaching the constitution. However, it did uphold citizens and future generations' constitutional right to a healthy environment. In February, the plaintiffs decided to take their legal battle to the Supreme Court.

Advancing an anticipatory approach to climate justice

What is clear from the analysis so far is the increasingly active role played by courts across the world in opening up new pathways of communication between publics and state agencies, as well as accountability for poor regulatory practices. The signing of the Paris Agreement in December 2015 marked a significant turning point in states' expressive commitment to meet emissions targets and to establish a rigorous system of accountability to ensure they do so (ibid.: Article 13). In the period since, the threat of legal prosecution for failing to meet targets has become very real. As a consequence, law is seen as providing publics with a stronger range of legal instruments with which to hold states accountable for not protecting their interests and fulfilling duties to secure a safe and sustainable future. For youth campaigners especially, law appears to offer a platform with which to advance their 'struggle for recognition' (Honneth, 1996), compensating somewhat for their lack of institutionalized political power (lack of voting rights) and countering efforts to diminish their credibility as legitimate sources of insight on climate change harms. Youth-led legal cases mobilize greater public interest in the performance of states and intensify pressure on government agencies to recognize international obligations to meet agreed emissions targets. Julia Olson, executive director of Our Children's Trust (2016) describes how 'youth are rising up globally and taking their governments to court to seek protection of their inalienable rights to a stable climate system Olson, Executive Director of Our Children's Trust, 2016).[3]

In doing so, mobilized youth draw wider attention to their civil rights to participate in processes of democratic decision-making (e.g., Convention on the rights of the Child, 1989, Article 12; Article 6(a) of the United Nations Framework Convention on Climate Change; Article 12 of the Paris Agreement), as well as their social rights to a fair distribution of essential resources (Our Children's Trust, Earth Guardians & The Global Initiative, 2016). These campaigners highlight how a logic of exclusion has worked to silence voices of dissent and to prevent a fuller recognition of the rights of youth to a safe and healthy future. Youth draw heavily on both state constitutional commitments, as well as international legislation, in their struggle to achieve a stronger recognition of their rights. In doing so, youth bring greater public attention to the relational character of justice and the need for all peoples to be afforded the opportunity to clarify their grievances and rights position. The greater communicative freedom afforded by legal proceedings has allowed youth campaigners to advance beyond a purely reactionary position, a more anticipatory approach to climate justice, showing how manageable thresholds of CO_2 emissions are possible, as is the preservation of necessary reserves of essential resources for the future. The understanding is that ongoing responsible environmental protection is not only practical but morally and legally required if practices that knowingly produce unjust outcomes are to be transformed.

From the start, the efforts of the international community to address climate change have recognized the importance of cooperation. From the creation of the Intergovernmental Panel on Climate Change in 1988, the adoption of the United Nations Framework Convention on Climate Change in 1992, to the signing of the Paris Agreement in 2015, states have consistently lent their support (in terms of legal agreements) to the idea of 'joint and separate action' to address climate change. There is, however, a certain compulsory element to this cooperation that should be noted. Since the Paris Agreement (2015), efforts to address climate change have come to be framed more explicitly as efforts to protect human rights extraterritorially, that is, through 'international assistance and cooperation' (see also the Maastricht Principles on Extraterritorial Obligations of States in the Area of Economic, Social and Cultural Rights, 2011, and the Vienna Declaration and Programme of Action (1993), Article 2, which requires states to engage in human rights cooperation). As the most 'paradigmatic example of global threat' today, climate change cannot be addressed effectively without internationally coordinated action and a stronger commitment from all to meet emissions targets (see Human Rights Council Report, A/HRC/31/52; Also, Aarhus Convention on Access to Information, Public Participation in Decision Making and Access to Justice in Environmental Matters, 1998). President Trump's (2017) decision to withdraw the United States from the Paris Agreement is a glaring violation of principles of international cooperation and contrary to what the UNFCCC (1992) defines as an 'appropriate international response' to deteriorating climate conditions. The danger is that the Trump administration will set a precedent for other larger polluting states to follow (Davenport, 2017). Solidarity with an inclusive model of climate mitigation is not only the essential component for the achievement of real progress on this front. It is also one that, in principle, could be insisted upon (given states' legal commitments to international cooperation), especially as ecological conditions deteriorate further globally.

Holding all states legally accountable for environmentally destructive practices is the primary aim of youth-led coalitions and their legal representatives at present. The realization, however, is that addressing the problems created by climate change will not be achieved through a series of court battles alone. Climate change mitigation requires an ongoing, renewable effort supported by a democratic culture that encourages human rights norms to be continuously reexamined in light of a range of problem areas. It also requires a progressive elaboration of the 'universal' dimensions of democratic futures, inclusive of the voices of those who, until now, have been largely ignored.[4] Mobilized youth seize the language of rights and make apparent the glaring contradictions that have emerged between the facticity of deteriorating environmental conditions and legally relevant notions of the inclusive democratic society. Youth insist that human rights discourse speaks to the lived circumstances of their lives, those who are direct witnesses to deepening ecological destruction (e.g., victims of disaster) and to expert knowledge, all of which draw attention to the inadequacies of states' current efforts to control pollution levels. The fact that courts are more willing today to upgrade 'interests' in shared resources (rivers, oceans, the atmosphere, etc.) to the status of a right signals something about changes occurring more generally in societal reasoning about climate change. First, it suggests that a greater clarity is beginning to emerge on the long-term goal of legal responsibilities to protect humanity's future. Second, a willingness seems evident to support a 'poly-contextualization' (Teubner, 2003) of human rights law, where the latter's relevance is explored more rigorously across a range of issue areas and social settings (Arato, 1998; McCarthy, 2004), including those linked to the future. Legal challenges help to further this process of human rights elaboration at the international, national, and regional levels, generating newer insights on how ideas of the just society can be resituated within settings marked by unprecedented environmental challenges.

Notes

1 See full verdict at: http://uitspraken.rechtspraak.nl/inziendocument?id=ECLI:NL:RB DHA:2015:7196.
2 Nature and Youth is an independent youth organization with 7,672 members throughout Norway. According to its Articles of Association, the organization is committed to protecting future generations who are 'dependent on the preservation of the natural environment's functions, productivity and diversity'.

3 Our Children's Trust is a nonprofit organization campaigning with those who have 'most to lose in the climate crisis' and to secure 'the legal right to a healthy atmosphere and stable climate on behalf of present and future generations' (see also Conca, 2014). This actor utilizes legal expertise to pressurize governments to reduce atmospheric carbon emissions.

4 Apart from youth, those whose interests are largely ignored are the poorer communities of the developing world living in disaster-prone regions whose contribution to climate change tends to be minimal yet which shoulder a disproportionate amount of its burdens. According to the IPCC (AR4 WGII Report: 275), an additional 600 million people face malnutrition due to climate change, with a particularly negative effect on Sub-Saharan Africa. Representatives of both climate-vulnerable communities and youth campaigners call on governments to give operational meaning to a principle of equity 'in accordance with their common but differentiated responsibilities and respective capabilities' (UNF-CCC, Article 3, paras. 1 and 2). Both defend the right of peoples not to be deprived of their means of subsistence, yet this and a series of related rights (to food and water security, for instance) are threatened by rising sea levels, severe drought and extreme weather events (Annual Report of the United Nations High Commissioner for Human Rights and Reports of the Office of the High Commissioner and the Secretary General, 2009).

References

Aarhus Convention on Access to Information, Public Participation in Decision-Making and Access to Justice in Environmental Matters (1998) available at www.unece.org/env/pp/treatytext.html (accessed: 27 August 2017).

Arato, Andrew (1998) 'Procedural law and civil society: Interpreting the radical democratic paradigm', in Rosenfeld, M. & Arato, A. (eds.) *Habermas on Law and Democracy: Critical Exchanges*, Berkeley: University of California Press: 26–36.

Barroux, Rémi (2016) 'Quel est le contexte juridique du vrai-faux "procès" de Monsanto?' *Le Monde* (16 October) available at www.lemonde.fr/planete/article/2016/10/16/nous-allons-verifier-si-les-activites-de-monsanto-sont-en-conformite-avec-les-regles-de-droit_5014553_3244.html (accessed: 24 March 2017).

Blumm, Michael C. & Guthrie, Rachel (2012) 'Internationalizing the public trust doctrine: Natural law and constitutional and statutory approaches to fulfilling the Saxion vision', The University of California Press, Davis, 45: 741–808, available at https://megward-wf6y.squarespace.com/s/2012_Blumm_Internationalizing-PTD.pdf (accessed: 24 August 2017).

Blumm, Michael C. & Wood, Mary C. (2013) *The Public Trust Doctrine in Environmental and Natural Resource Law*, Durham, NC: Carolina Academic Press.

Colorado Court of Appeals (2017) 'Case No. 16CA0564' (23rd March) available at https://static1.squarespace.com/static/571d109b04426270152febe0/t/58d4170fbe65943deef84303/1490294544055/2017.03.23+Martinez++Opinion.pdf (accessed: 23 March 2017).

Conca, James (2014) 'Atmospheric trust litigation – can we sue ourselves over climate change?' *Forbes* (23 November) available at www.forbes.com/sites/jamesconca/2014/11/23/atmospheric-trust-litigation-can-we-sue-ourselves-over-climate-change/ (accessed: 8 April 2017).

Conca, James (2016) 'Federal court rules on climate change in favor of today's children', *Forbes*, available at www.forbes.com/sites/jamesconca/2016/04/10/federal-court-rules-on-climate-change-in-favor-of-todays-children/ (accessed: 23 March 2017).

Cornell, Drucilla (1990) 'The violence of the masquerade: Law dressed up as justice', *Cardozo Law Review* 11: 1047–1064.

Davenport, Coral (2017) 'Top Trump Advisers are split on Paris Agreement on climate change', *New York Times*, 2 March.

Earth Guardians and Our Children's Trust (2014) 'Colorado oil and gas conservation commission denies young people the right to a healthy future' (4 June) available at https://static1.squarespace.com/static/571d109b04426270152febe0/t/5760d21e1d07c0ae983539d7/1465963039171/14.06.04COPetition.pdf (accessed: 24 March 2017).

Eradicating Ecocide (2015) 'Fact sheet' available at http://www/eradicatingecocide.com/wp-content/uploads/2015/11/EL-factsheet_English11.15.pdf (accessed: 24 March 2017).

European Environmental Agency (2016) 'Air quality in Europe – 2016 report' available at www.eea.europa.eu/publications/air-quality-in-europe-2016/download (accessed: 3 July 2018).

Gosseries, Axel (2008) 'Future generations' future rights', *The Journal of Political Philosophy*, 16(4): 446–474.

Government of Pakistan National Climate Change Policy (2012) available at http://www.gcisc.org.pk/National_Climate_Change_Policy_2012.pdf (accessed: 7 December 2018).

Greenpeace International (2017) 'Climate lawsuit against Arctic oil goes to court', 15 February, available at www.greenpeace.org/international/en/press/releases/2017/Climate-lawsuit-against-Arctic-oil-goes-to-court/ (accessed: 15 June 2017).

Honneth, Axel (1996) *The Struggle for Recognition*, trans. J. Anderson, Cambridge, MA: MIT Press.

Human Rights Council (2016) 'Report of the special rapporteur on the issue of human rights obligations relating to the enjoyment of a safe, clean, healthy and sustainable environment' (1 February) (A/HRC/31/52) available at https://papers.ssrn.com/sol3/papers.cfm?abstract_id=314845 (accessed: 5 July 2018).

Human Rights Council (2017) 'Human rights and climate change' (22 June) (A/HCR/35/20) available at www.right-docs.org/doc/a-hrc-res-35-20/ (accessed: 5 July 2018).

Indian Supreme Court (1996) '*M.C. Mehta v. Kamal Nath & Ors*' available at https://indiankanoon.org/doc/1514672/ (accessed: 3 July 2018).

International Covenant on Economic, Social & Cultural Rights (1966) available at www.ohchr.org/EN/ProfessionalInterest/Pages/CESCR.aspx (accessed: 3 July 2018).

Jefferson, Thomas (1905) 'Letter to James Madison' [Paris, 1789], in *The Works of Thomas Jefferson, Vol. 6 (Correspondence 1789–1792)*, New York and London: G. P. Putnam's Sons/ Knickerbocker Press: 3–11.

Just Planet (2016) ' "States" obligations to stabilize the climate for the life, survival and development of children and future generations' (23 September) available at www.ohchr.org/Documents/HRBodies/CRC/Discussions/2016/JustPlanet.pdf (accessed: 27 August 2017).

Knox, John H. (2009) 'Linking human rights and climate change at the United Nations', *Harvard Environmental Law Review*, 33(2): 477–498.

Labrador Inuit Ass'n v. Nfld. (Minister of Env't and Labour) (1997) '155 Nfld. & P.E.I.R. 93 (Can.)' available at https://static1.squarespace.com/static/.../t/.../2012_Blumm_Internationalizing-PTD.pdf (accessed: 3 July 2018).

Lahore High Court Green Bench (2015) *Asghar Leghari v. Federation of Pakistan (W.P. No. 25501/2015)* available at https://elaw.org/pk_Leghari (accessed: 24 March 2017).

The Lancet Commission (2017) 'The Lancet commission on pollution and health' available at www.thelancet.com/commissions/pollution-and-health (accessed: 8 November 2017).

Locke, John (1988) *Two Treaties of Government*, ed. Peter Laslett, Cambridge: Cambridge University Press.

Maastricht Principles on Extraterritorial Obligations of States in the Area of Economic, Social and Cultural Rights (2011) available at www.etoconsortium.org/nc/en/main-navigation/library/maastricht-principles/?tx_drblob_pi1%5Bdow (accessed: 25 August 2017).

McCarthy, Thomas (2004) 'Political philosophy and racial injustice: From normative to critical theory', in Benhabib, Seyla & Fraser, Nancy (eds.) *Pragmatism, Critique, Judgment*, MIT Press: 147–168.

Moloney, Anastasia (2018) 'Colombia's top court orders government to protect Amazon forest in landmark case', *Reuters*, 6 April.

Monsanto (2016) 'An open letter about the Monsanto Tribunal' available at https://mon santoblog.eu/an-open-letter-about-the-monsanto-tribunal/ (accessed 7 December 2018).

Nature and Youth – Young Friends of the Earth Norway (2017) 'Articles of association' available at https://nu.no/english/ (accessed 7 December 218).

Neslen, Arthur (2016) 'Norway faces climate lawsuit over Arctic oil exploration plans', *The Guardian*, 18 October.

Norwegian Constitution (1814) available at https://ihl-databases.icrc.org/ihl-nat/6fa4d35e 5e3025394125673e00508143/eee956c813a2da0ec1256a870049de0c/%24FILE/Consti tution.pdf (accessed 7 December 2018).

Office of the High Commissioner for Human Rights (2015) 'Understanding human rights and climate change' available at www.ohchr.org/Documents/Issues/ClimateChange/ COP21.pdf (accessed: 23 August 2017).

Our Children's Trust (2016) 'Youth file lawsuit against Norwegian government over Arctic oil' (18 October) available at https://static1.squarespace.com/static/571d109b 04426270152febe0/t/5807bba8579fb39e1dd6639f/1476901803100/NorwayWritSum mons.pdf (accessed: 23 March 2016).

Our Children's Trust (2017) 'In *Youth v. Colorado Oil & Gas*, API – youth win' available at https://meg-ward-wf6y.squarespace.com/s/170323-CO-Court-of-Appeals-Decision-PR.pdf (accessed: 3 July 2018).

Our Children's Trust, Earth Guardians & The Global Initiative (2016) 'State obligations regarding Children's rights and climate change', Submission to UN Committee on the Rights of the Child' available at www.ourchildrenstrust.org/s/OCT-et-al-CRC-Submis sion.pdf (accessed: 8 April 2017).

Peterson, Merrill (1976) 'Mr. Jefferson's sovereignty of the living generation', *Virginia Quarterly Review* (Summer Edition): 437–447.

Republic of the Philippines Supreme Court (1993) 'G. R. No. 101083 30 July, *Minors Oposa v. Secretary of the Department of Environment and Natural Resources (DENR)*' available at https://www.cambridge.org/core/journals/international-legal-materials/ article/philippines-supreme-court-decision-in-minors-oposa-v-secretary-of-the-depart ment-of-environment-and-natural-resources-denr-deforestation-environmental-dam age-intergenerational-equity/1D346EABA84AA86E9FD855C2A83ABB31 (accessed: 20 March 2017).

Superior Court of the State of Washington for King County (2015) 'Ruling of Judge Hollis R. Hill on *Zoe and Stella Foster v. Washington Department of Ecology*' (19 Novem-ber) available at https://static1.squarespace.com/static/571d109b04426270152febe0/t/

57607fe459827eb8741a852c/1465941993492/15.11.19.Order_FosterV.Ecology.pdf (accessed: 24 March 2017).

Superior Court of the State of Washington in and for the County of King (2016) *Zoe & Stella Foster et al. v. Washington Department of Ecology* (19 December) available at https://static1.squarespace.com/static/571d109b04426270152febe0/t/585979e1d1758e c9d1667705/1482343090836/FostervEcology-2016-12-19-141247 (accessed: 24 March 2017).

Teubner, Gunther (2003) 'Regime collisions: The vain search for legal unity in the interpretation of global law', *Michigan Journal of International Law*, 25(999): 1005–1017.

Touraine, Alain (1988) *Return of the Actor: Social Theory in Postindustrial Society*, Minneapolis: University of Minnesota Press.

United Nations Convention on the Rights of the Child (1989) available at www.unicef.org. uk/what-we-do/un-convention-child-rights/ (accessed: 24 March 2017).

United Nations Environment Programme (2015) 'Annual Report 2015' available at www. unenvironment.org/annualreport/2015/en/index.html (accessed: 7 December 2018).

United Nations Framework Convention on Climate Change (1992) available at https:// unfccc.int/resource/docs/convkp/conveng.pdf (accessed: 3 July 2018).

United Nations Office of the High Commissioner for Human Rights (2016) 'OHCHR's key messages on human rights: Climate Change and Migration' available at www.ohchr.org/ Documents/Issues/ClimateChange/Key_Messages_HR_CC_Migration.pdf (accessed: 5 July 2018).

United Nations Paris Agreement (2015) available at www.un.org/sustainabledevelopment/ climatechange/ (accessed: 8 April 2017).

United States Federal District Court in Eugene, Oregon Complaint for Declaratory and Injunctive Relief at 85, *Kelsey Cascadia Rose Juliana, Xiuhtezcatl Tonatiuh M. Et Al. v. United States, Barack Obama et al.*, No. 6:15-cv-01517-TC (2015) available at www.supremecourt.gov/DocketPDF/18/18-505/67251/20181017183026537_In%20 re%20United%20States%20%20-%20Petition%20for%20Mandamus.pdf (accessed: 6 December 2018).

United States Supreme Court, Illinois Central Railroad v. Illinois, 146 U.S. 387 (1892) available at https://supreme.justia.com/cases/federal/us/146/387/ (accessed: 3 July 2018).

Vienna Declaration and Programme of Action (1993) available at www.ohchr.org/Docu ments/ProfessionalInterest/vienna.pdf (accessed: 3 July 2018).

Washington Environmental Council v. Bellon (2013) 'US court of appeals for the ninth circuit' available at https://caselaw.findlaw.com/us-9th-circuit/1646826.html (accessed: 3 July 2018).

Wijdekop, Femke (2016) 'The duty to care for our common home', *New Internationalist*, available at www.eradicatingecocide.com/wp-content/uploads/2012/06/NewInt_Femke_ May16.pdf (accessed: 16 March 2017).

Woods, Mary Christina (2014a) 'Beyond borders: Shared ecology and the duties of Sovereign Co-tenant trustees', in *Nature's Trust: Environmental Law for a New Ecological Age*, Cambridge: Cambridge University Press.

Woods, Mary Christina (2014b) 'Tribal trustees in climate crisis', *American Indian Law Journal*, 2(2) (Spring): 518–546.

World Economic Forum (2016) 'Rising pollution in the developing world' available at http://reports.weforum.org/outlook-global-agenda-2015/top-10-trends-of-2015/6-rising-pollution-in-the-developing-world/ (accessed: 24 August 2017).

World Future Council (2013) 'Crimes against future generations' available at www.world futurecouncil.org/crimes-against-future-generations/ (accessed: 3 July 2018).

Young, Iris Marion (2011) *Responsibility for Justice*, Oxford: Oxford University Press.

Zoe and Stella Foster, et al. v. Washington Department of Ecology (2015) available at www. crin.org/en/library/legal-database/zoe-and-stella-foster-et-al-v-washington-department-ecology (accessed: 7 December 2018).

Chapter 5

Publicly embedded constitutions

Legislating for present and future generations

Introduction

The type of legal accomplishments noted in Chapter 4 emerge because of one fundamental fact that is sometimes overlooked by social movement research: the constitutional identity of the state as a defender of democratic justice (e.g., Edwards & Gillham, 2013).[1] It is this identity that creates certain opportunities for publics to challenge the legitimacy of the state's ongoing commitments to fair representation (Blokker, 2013: 68) and to influence the shape of future environmental justice. These powers derive from the status of being a citizen of a state that claims that its democratic project is collectively achieved by 'we the peoples' and their political representatives. The understanding, therefore, is that state constitutional law offers its publics an important resource with which to address environmental problems. There is no denying the extent to which states today are guilty of negligence in environmental pollution control. However, these same states also preserve the capacity to initiate necessary changes to address this problem, a fact defended both within state sovereign borders (by citizen campaigners) and simultaneously beyond in wider international settings (e.g., UNFCCC, 1992; Habermas, 2008: 444). The state continues to provide the core prerequisites for the exercise of democratic self-determination across a range of issues and, in that, holds out immense corrective capacity to deal with ecological problems of one kind or another. Thus, whilst the rule of law in a modern democratic state system is not premised on an absence of violence against the natural world, it, nonetheless, does reflect the 'constitutional spirit' of peoples who regularly challenge the legitimacy of such violence from the perspective of constitutional commitments to environmental protection, rights to health, life and liberty. Campaigners note the importance of historical achievements in this regard at the same time as they criticize one-sided applications of justice ideals. Given current levels of legal mobilization on climate change issues, the likelihood is that this debate will continue to develop further in the future. Democracy is not, as Marx (1852) once wrote, the 'solved riddle of all constitutions'. As ecological problems deepen, climate change is likely to further confound 'the riddle' of state constitutional commitments to equality, freedom and democratic solidarity (Marks, 2000). Indeed, this 'riddle' must always remain open to the interpretive efforts of democratic publics (Brunkhorst, 2012) as they struggle to make sense of justice commitments in a climate-challenged world.

This chapter explores the nature of these commitments as they are expressed as state constitutional rules aimed at protecting the health of the environment for present and future generations (e.g., Constitution of Norway, Article 110b; Constitution of Colombia, Article 79, 80, 88). It further considers whether these principles offer a legitimate basis for the initiation of a stronger institutional commitment to environmental security as an intergenerational public right. It notes how these commitments are articulated by various constitutional states as culturally specific interpretations of what is means to be a trustee of that state's democratic, cultural and ecological heritage, as well as universal justice ideals (e.g., human rights, civil rights, etc.). Commitments to intergenerational environmental justice emerge from the efforts of various state polities to interpret established principles of justice and legal political traditions in light of new ecological challenges. Practical judgments as to what is fair and just regularly reassess the validity of these legal commitments against the lived circumstances of community members (Buchwalter, 1998: 3). When the full validity of democratic constitutional rules is felt to be unrealized (as has regularly been the case over the past centuries), citizens mobilize to change the law. For example, when Thomas Jefferson (1776. ME 1:29, Papers 1:315) wrote about 'the inherent and inalienable rights' of all people, at the time his ideas did not, in fact, extend to all peoples (see Hunt, 2007: 18). With time, interpretations of valid recipients of inalienable rights evolved, as various 'forgotten' categories of peoples (women, slaves, religious minorities children, the disabled, ethnic minorities, indigenous peoples, etc.) came to be included in official legal interpretation. However much a legal constitutional order may stress its autonomy from the circumstances of everyday life, it, nonetheless, preserves an important connection to the same via the state's citizenry who mobilize to change the law when it does not offer due recognition of their interests.

Constitutional law may rely on certain established traditions of knowledge and ready-made mechanisms of justification that ground the validity of its rules in a system of communication that seems to 'turn back upon itself' to preserve the authority of its position (Luhmann, 1987, 1997). However, democratic constitutional law, by its very nature, always preserves the relevance of its schema of interpretation by keeping the meaning and act of self-determination open to new possibilities, in response to various societal challenges, including climate change. What is 'rightful' and 'just' is thereby also determined by law's engagement with the social (Lenoble & Maesschalck, 2016: 81). Ideas and principles of intergenerational justice are debated socially and relevant legal categories reassessed in terms of newly emerging ecological realities. The 'civic track' (Teubner, 1992) that connects constitutional law with its publics allows for a communication on the ongoing validity of its rules (as originally intended; see Kant, 1970[1784]: 57–58). In that, constitutional law offers publics an important tool with which to further elaborate the legal basis of environmental justice. The ecological welfare of living and future generations proves to be a serious driver of a new wave of legal challenges centering largely on the relevance of state constitutional law to a wide range of environmental concerns today (e.g., atmospheric justice). Accumulating threats to atmospheric safety and basic ecological well-being enhance the societal relevance of constitutional law as a protector of human rights, as well as the notion of sovereign democracy as belonging to the people in whose name the powers of government are exercised (Castiglione & Warren, 2006: 18). With the understanding that unmitigated climate change poses a serious threat to human rights to health, safety and development (e.g., see UN General Assembly, January 15, 2009) comes the realization that it also poses a threat to democracy (commitments to the social, economic and developmental well-being of 'we the peoples'). Constitutional law is increasingly called upon to clarify how the state ought to address deepening ecological problems with the welfare of present and future citizens firmly in mind.

The fact that constitutions are continuously tested, occasionally revised and new stipulations added illustrates how the certitude of democratic constitutional justice is never guaranteed or closed to further revision (Wolfrum & Grote, 2011). The legal cases explored in Chapter 4 of this book highlight the importance of constitutional law as a mobilization resource, one entirely underestimated by social movement research to date. Legal cases attract public and media attention but also force legal authorities and government agencies to reflect more deeply on the legitimacy of prevailing environmental practices. The 'surplus' of meanings generated through such interpretive exercises frequently filters into wider social and political discourse. One consequence of the these developments is the re-examination of the meaning of self-determination from a more peoples-centered perspective (McCarthy, 1994: 21). Similarly, the type of arguments presented through legal cases seeking clarification on the 'climate rights' of future others encourage an extension of democratic rights beyond established legal interpretation (Cornell, 1990: 267). Citizens assert the will to exercise their democratic powers in response to new threats to their ecological well-being and that of their children. On occasion, this leads to calls for the amendment of existing constitutional rules. Granting greater recognition to the welfare of future generations is not something 'foreign' to state democracies but, a natural process of maturation in collective moral reasoning, 'the becoming of ourselves and of our own knowledge' (Hegel, 1892: 4). Concern for future humanity has preoccupied thinkers for centuries (e.g., Jefferson [1789], Condorcet [1789] (Williams, 2004), Burke [1890]). Its pertinence today is driven home by the risk of ecological collapse and planetary-wide threats of destruction.

Societal discourse on environmental rights and relations of justice between generations intensifies as a consequence. In some instances, this discourse has already led to amendments to constitutional law. In other cases, a more heated debate on the ongoing relevance of existing legal provisions recognizing the right to health, development and environmental security ensues. A key concern is how a regime of justice that boasts equity, fairness and solidarity can justify its contribution to escalating ecological destruction. Not to respond actively to climate change as a major risk to future survival threatens the lives of billions. However, it also threatens to undermine core elements of democratic civic identity (states as protectors of their members interests; also Europe's self-proclaimed identity as 'a Europe that protects', European Commission, 2017) and delimit the transformative capacities of constitutional democratic communities to address major challenges, as they have done in the past (e.g., world wars, the threat of nuclear war, etc.).

When revolutionaries first criticized the idea of a perpetual law and declared living generations as 'not entitled to subject future generations to its own laws' (first French Constitution (1793), Article 28), they could not have foreseen the relevance of their thinking to contemporary times. Stipulations such as that claiming 'choices designed to meet the needs of the present generation should not jeopardize the ability of future generations and other peoples to meet their own needs' (French Constitution, Amendments in 2008, Charter for the Environment) are of little value if they do not have an impact on everyday social and environmental practices. One of the more important contributions of future and youth justice coalitions to climate justice debates to date has been their reassertion of the normative relevance of these constitutional principles. They also draw attention to a number of assumptions in liberal state democratic reasoning that have become somewhat problematic. For instance, the notions that newer generations will have an opportunity in the future to reshape the constitutional rules of their society once they reach voting age and further and that the conditions necessary to allow them to practice self-government will continue unhindered.

The main problem with this formulation of intergenerational justice is that it does not reflect the true nature of contemporary circumstances (the speed of escalating climate degradation, rising national debt, security threats, enduring poverty, etc.) that threaten newer generations' *capacity* to be self-governing. Taking just one of the many sources of threat today, rapidly diminishing supplies of essential resources necessitate immediate measures to ensure that a more rational distribution is secured across generations, in line with constitutional commitments. Replacing more traditional notions of an open-ended future (with equal opportunities enjoyed by each successive generation to shape the circumstances of their own lives) is that of an ecologically fragile 'present future', one that is deeply affected and already largely shaped by the cumulative pollution practices of previous generations. Ecologically precarious futures are futures that must be planned for, deliberated upon and actively protected in the here and now. According to the legal and moral convictions of most democratic communities, peoples of the future have recognizable interests that are under serious threat. Why then should institutional practices not reflect this concern? Extending an ethic of care to such peoples also requires a political consensus that protections will, in fact, be assigned to those not capable themselves at this point of making claims or defending their future resource interests. Moves to constitutionalize deeper relations of justice between generations and lower the voting age (to 16) are two changes considered by many as essential if more sustainable Anthropocene futures are to be realized. Youth claim that current constitutional rules are exclusionary on a number of fronts. Most notably, in not granting those under the age of 18 the right to vote, they are denied the opportunity to shape current policy choices (e.g., energy policy) that will profoundly affect their future. If we consider the real issue here to be that of self-government, the expressed dissatisfactions of soon to be voting cohorts is interesting because it raises the question of whether the diminishing time scales available to address grave ecological problems, in fact, make youth's accusations of exclusion and constitutional injustice valid? Before addressing this issue, the following analysis first considers how the interests and welfare of future generations are currently represented across a range of state constitutions. The examples discussed here reflect the will of sovereign peoples who choose (at least in principle) to make explicit their commitment to environmental justice across generations.

Existing state constitutional references to future generations

Many state constitutions refer to 'responsibilities towards future generations'. However, such a reference usually forms part of the interpretive background to the main principles of a constitution. To have real enabling force, references to future generations ideally should be embedded in the substantive part of a state's constitution where rights and duties are articulated in more detail. Several constitutions do include such provisions. Article 225 of Brazil's constitution, for instance, refers to the duties of government and the community 'to defend and preserve the environment for present and future generations'. Similarly, Article 20(a) of the German constitution provides that the 'state protects the natural living conditions also in responsibility for future generations' (Basic Law for the Federal Republic of Germany), while the constitution of South Africa expressly affirms everyone's 'right . . . to have the environment protected, for the benefit of present and future generations' (South African Constitution, Article 24). The constitution of Belgium obliges government to take future generations into account when enacting state functions (Belgium Constitution, Article 7b). The constitutions of Bhutan (Article 5) and Portugal (Article 66) refer to the fair allocation of burdens and benefits equally across generations. In environmental terms, this means ensuring fairness in the intertemporal distribution of natural resource endowments and their exploitation. Effective since January 2012, Hungary's new constitution refers to 'all natural resources, especially agricultural land, forests and drinking water supplies, biodiversity' as part of the nation's common heritage; 'the state and every person shall be obliged to protect, sustain and preserve them for future generations' (Article P). The Pennsylvania state constitution clarifies, amongst other things, how 'public natural resources are the common property of all the people, including generations yet to come' (Article 1).

Constitutions granting specific rights to future generations include that of the Plurinational State of Bolivia (Article 7), which clarifies all citizens' fundamental right 'to enjoy a healthy environment, ecologically well balanced and appropriate to her wellbeing, while keeping in mind the rights of future generations'. Article 11 (Chapter III) of Japan's constitution specifies how the 'fundamental human rights guaranteed to the people by this constitution shall be conferred upon the people of this and future generations as eternal and inviolable rights'. The constitution of Norway refers to the duties of the Norwegian people to 'maintain' each individual's 'right to an environment that is conducive to health and to a natural environment' (Article 112). Natural resources, it adds, must be 'made use of on the basis of comprehensive long-term considerations whereby this right will be safeguarded for future generations as well' (Article 110b). Norway's constitution, therefore, not only refers to the importance of an equitable distribution of resources across generations. It also offers a procedural protection of rights to the same, one that encompasses present and future generations.

Strong constitutional references to future generations, especially rights-based references, offer legal courts a substantial resource with which to initiate preventative actions and ensure accountability when rights are violated and responsibilities not fulfilled. For instance, the decision of Nature and Youth and Greenpeace Nordic to file a lawsuit against the Norwegian government in 2017 on the grounds that its decision (in 2016) to grant new oil extraction licenses in areas of the Arctic Barents Sea with no existing petroleum infrastructure was a violation of citizens' constitutional protections (e.g., the right of citizens to a safe and healthy environment and the rights of future generations to the same, as well as the rights of citizens to be 'informed of the state of the natural environment and of the effects of any encroachments on nature that are planned or commenced' (Article 110b)). The coplaintiffs describe their decision to launch this case as driven by their sense of 'duty as citizens' to act when politicians fail to fulfill constitutional commitments, as well as responsibilities in terms of international law (e.g., terms of the Paris Agreement, 2015): 'The climate lawsuit is driven by people [who wish] to hold the Norwegian government accountable under law to ensure a healthy and safe environment and climate for us and for future generations' (see Save the Arctic, 2017). Although the Oslo District Court ruled in favor of the Norwegian government in January 2018, it did, however, acknowledge the plaintiffs right to defend 'citizens and future generations constitutional right to a healthy environment' (Oslo District Court Proceedings, Case No.: 16–166674TVI-OTIR/06,: 16). In early February 2018, Nature and Youth and Greenpeace (with the additional support of *Grandparents Climate Campaign*) decided to appeal the decision of the court and take this legal battle to the Supreme Court.

Beyond their purely legal function, therefore, constitutional references to future generations also give expression to citizens' self-understandings as a sovereign peoples, that is, as a people connected to and protective of posterity. The 'who' of justice is understood as also extending firmly beyond the present and relevant parties to the democratic process, therefore, encompass future others. At very least, constitutional references to future generations increase the likelihood of a more equitable distribution of essential resources across time.

Several states have taken it upon themselves to reinforce constitutional commitments to protect the welfare of future peoples by establishing a new Ombuds Office or Commissioner for Future Generations. The first proposal to establish a 'guardian for future generations' came in March 1992 when the Maltese government made a contribution to the preparatory meeting of the United Nations Conference on Environment and Development in Rio, outlining how a new representative body was needed to alert the wider international community to threats to the well-being of future generations. In the years since, National Parliamentary Commissions for Future Generations, for the Environment, for Natural Resources and/or Sustainable Development have been established in several states, chiefly to kick-start a more effective means of representing the interests of future generations at a practical policy level. For instance, in 2007 the Hungarian Parliament established a Parliamentary Commissioner for Future Generations. This was one of three Ombuds offices responsible for the protection of specific rights, the other two being The Parliamentary Commissioner for Civil Rights and the Parliamentary Commissioner for the Rights of National and Ethnic Minorities. The primary task of Hungary's Ombuds for Future Generations was 'to ensure the protection of the fundamental right to a healthy environment' and to enable 'with attention, estimate, and control' the emergence of provisions in law that ensure 'the sustainability and improvement' of the environment for all (Hungarian Act on Parliamentary Commissioner for Future Generations, 27/B (1); see Ambrusné, 2010). This office was first proposed by the civil society organization Védegylet (Protect the Future), and its success throughout its years in operation is perhaps best explained by the fact that it started as a popular grassroots initiative (Göpel, 2010; World Future Council, 2011).

Hungary's first appointed Commissioner for Future Generations, Sándor Fülöp (May 2008–August 2012), describes how this office was, from the start, committed to responding to citizens' concerns over specific development projects and launching independent investigations when appropriate. Fülöp explains how this office performed both an advocacy and an investigative role, placing the right to a clean and healthy environment above all other concerns and monitoring policy developments to ensure that the latter did not pose a threat to the welfare of future sovereign communities. In the first 14 months of its operation, the commissioner's office received 600 complaints from concerned citizens (Johnson & Asquith, 2010). Information was regularly exchanged with local citizen groups, a form of public engagement known as 'crowd sourcing', to encourage publics to report on regulatory 'flaws', as they appeared. These were subsequently investigated by the Parliamentary Commissioner's office. With the adoption of a new Basic Law, the Ombuds offices were merged and in January 2012, replaced with a unified office known as the Office of the Commissioner for Fundamental Rights, exercising 'soft powers' of investigation of violations of the fundamental rights of vulnerable peoples. This new office can make recommendations to relevant authorities regarding best practice, but its decisions are not legally binding. The Commissioner for Fundamental Rights has two deputies: the Deputy Commissioner for

Fundamental Rights responsible for the protection of the rights of national minorities and the Deputy Commissioner for Fundamental Rights responsible for the protection of the interests of future generations.

In Israel, a similar office was established in March 2001. Israel's Commission for Future Generations operated as an independent unit of parliament, investigating disputes over natural resources, education, health care policy or any other matter of special concern. Its aim, until its demise in 2006, was to ensure 'a dimension of the future' be included in all policy legislation. Similar to the situation in Hungary, Israel's Commissioner, Shlomo Shoham, had the authority to provide the Knesset (Israeli Parliament) with recommendations on any issue the Commissioner considered relevant to the welfare of future generations. The Commissioner also had the power to demand information from institutions 'subject to inspection by the State Controller' (including ministries, state corporations and local authorities). Göpel (2010) debates whether its closure in 2006 was due to the rigorous approach of Shoham in the conduct of the commission's affairs. Its demise was considered a blow to wider efforts encouraging states to establish similar offices and pioneer new policy approaches to intergenerational justice. In New Zealand, a Parliamentary Commissioner for the Environment was established in 1986, accountable to New Zealand's Parliament rather than the government of the day. This office continues to be concerned with improving approaches to environmental governance by investigating 'the effectiveness of environmental planning and management' on the part of public authorities, obtaining information on matters relating to emissions trading schemes and water allocation and offering advice on remedial action. The cumulative efforts of the head of this unit, Commissioner Jan Wright, have been praised by various sources, including Māori indigenous leaders, especially her willingness to acknowledge and take on board the views of indigenous peoples (Göpel, 2010).

In Germany, the Parliamentary Advisory Council on Sustainable Development, created in 2004, monitors and supports the federal government's sustainability policy at the national, EU and international levels by conducting sustainability impact assessments and reporting its findings to parliament. Following the recommendations of the United Nations Conference on Environment and Development at the Earth Summit in Rio de Janeiro in 1992, the Council recognizes the right to sustainable development as a global one, reaching across borders and generations. For instance, the Council affirms its commitment to ensuring 'that life today is not at the expense of tomorrow' and highlights the importance of the future to the decision making of the German Bundestag. In June 2016, the Council called for the establishment of four Regional Sustainability Networks in Germany, to be supported and coordinated by an office at the Council for Sustainable Development and for the principle of sustainability to be included in the German Constitution (Parliamentary Advisory Council on Sustainable Development, August 2016). In Wales, the Welsh Commissioner for Future Generations was established by the Welsh Assembly following the closure of the Sustainable Development Commission, an administrative body covering the whole of the United Kingdom, by the

UK government in 2011. The government at the time decided it was not appropriate for a nonelected body to offer advice to government on environmental matters. In 2014, new legislation was proposed in the Welsh Assembly, known as the 'Well-Being of Future Generations (Wales) Act (2015), which saw the introduction of a series of new sustainable development goals for public bodies and the establishment of a Future Generations Commissioner for Wales. The main duties of the new commissioner are to advise, assist and support public bodies regarding the attainment of sustainable welfare goals, as well as to investigate the long-term impact of their activities on future generations' interests (National Assembly for Wales, 2015). Public bodies are now obliged to set six 'well-being goals' and, in all their operations, to 'take account of the importance of balancing short-term needs' with those of the long-term. Promoted initially under the banner 'The Wales We Want' (in line with the UN's 'The World We Want' initiative, 2015), the appointment of a Future Generations Commissioner for Wales has proven particularly beneficial to efforts to improve public awareness of the need to protect the interests of future generations.

Efforts to establish independent offices for future generations in several other states and/or regions suggest that intergenerational justice imperatives are becoming more prominent at the policy level and some administrative offices more supportive of development indictors beyond GDP. More generally, there would appear to be greater awareness amongst publics as to the need to impose a series of checks and balances on structural short-term thinking (Göpel, 2010). The positive contribution of many regional and local offices to date has encouraged citizens to see themselves as part of a 'temporal series of sovereigns', where current voting publics represent future sovereigns by acting as trustees of the democratic process and ensuring that environmental security remains a collectively determined good shared across generations (Thompson, 2005: 248). Many campaign also at the European level for the establishment of an EU Ombuds Office for Future Generations (e.g., see Nesbit & Illes, 2015). Campaigners point to the Treaty of the European Union and its references to the need for 'solidarity between generations', as well as Article 3 of the Lisbon Treaty obliging Europe 'to promote peace' and 'the well-being' of 'all its peoples', including future generations (Bruch et al., 2002). Actors, including the Foundation for Democracy and Sustainable Development (2015: 4), argue that this legislation provides a strong justification for the establishment of a new EU Commissioner of Guardian for Future Generations to alert policy makers across Europe of changes needed to improve the efficacy and contribution of policy drafted in single-issue departments to sustainable development goals.

Not only would such a move signal a deeper collective endorsement of the European Union's principle of solidarity, it would also secure the permanence of a body capable of performing a watchdog and advisory role. The legitimacy of this office's policy interventions and communications with member states, however, would have to be secured in advance, as would the provision of reports outlining what changes various sectors would have to introduce in response to its recommendations. At present, there is insufficient support for the establishment of a new EU Guardian for Future Generations. However, that is not to say that such a body will not emerge in the years ahead, especially as public support for a more long-term and integrated policy approach to intergenerational justice continues to grow. The gradual realization is that more rigorous efforts are needed not only at the regional or local level but at the EU level in order to parallel efforts being made elsewhere.

Protecting future generations' future rights to self-determination

The first formal recognition, at the international level, of the right of future generations to a healthy environment emerged with the signing of the Stockholm Declaration in 1972. This document recognizes all members of the community of humanity's fundamental rights to 'a life of dignity and well-being' (Stockholm Declaration [Declaration of the United Nations Conference on the Human Environment], 1972, UN Doc. A/Conf.48/14/Rev.1). Principle 1 expresses the common conviction that humanity bears a solemn responsibility to protect and improve the environment for present and future generations. At least three objectives guide its interpretation of intergenerational justice. The first objective is to ensure that sufficient natural resource options are conserved to sustain a healthy planet into the future. The second is to conserve quality of life for all across time, and the third is to establish a minimum degree of equity in the distribution of scarce resource supplies. In relation to all three objectives, the project of justice is clearly failing, as global carbon emissions levels soar, sea levels rise and extreme weather events become more frequent and catastrophic in their effects (see UN World Meteorological Organization, 2016). A major commitment to action is needed if newer generations are to inherit a planet capable of meeting their basic resource needs. The issue of intergenerational justice was already a concern for the UN when it published *Our Common Future* in 1987, but by the time the United Nations Conference on Environment and Development (UNCED) was staged in 1992, it had become a more urgent concern. The Rio Declaration describes the right to development as one that must be fulfilled equitably so as to meet the 'developmental and environmental needs of present and future generations'. On account of its legally binding nature, the UNFCCC's references to future generations are more consequential. For instance, Article 3 directs all parties to 'protect the climate system for the benefit of present and future generations of humankind, on the basis of equity and in accordance with their common but differentiated responsibilities and respective capabilities'.

Other relevant legislation referencing future generations includes the United Nations Convention on the Law of the Sea (1982), the UN Convention to Combat Desertification (1994) and the Joint Convention on the Safety of Spent Fuel Management and on the Safety of Radioactive Waste Management (1997). At the World Summit for Social Development in 1995, states committed themselves to creating a framework of action that, amongst other things, ensures equity among generations by protecting the integrity and sustainability of resources into the future (Report of the World Summit for Social Development, 1995, Chapter One, Resolution 1, annex 1, para. 26[b]). In 2012 at the United Nations Conference on Sustainable Development (also referred to as Rio+20), parties endorsed the promotion of an 'economically, socially, and environmentally sustainable future' for 'present and future generations' (The Future We Want, Annex 1, *Our Common Vision*, 2012). In 2014, the UN General Assembly's Open Working Group on Sustainable Goals published its proposals for a set of new objectives to replace the Millennium Development Goals (which expired at the end of 2015). These included the eradication of poverty and hunger, stronger measures to promote health and climate change mitigation. All were affirmed as issues of intergenerational, as much as intragenerational justice. Given the tendency for poverty to be transmitted from parent to child (World Hunger and Poverty Facts and Statistics, 2015) and the burdens of climate change to be passed from present to future generations, sustainable goals for the international community must acknowledge the intertemporal nature of these global problems.

In August 2013, the office of the UN Secretary General presented its report 'Intergenerational Solidarity and the Needs of Future Generations' to the UN General Assembly as part of a wider initiative for the further implementation of Agenda 21 and the United Nations Conference on Sustainable Development. This report notes how intergenerational solidarity is already embedded in existing treaties, declarations, resolutions and intergovernmental decisions. One of the tasks of the international community, it argues, is to consider how it can best operationalize 'already existing' commitments to intergenerational solidarity. In terms of ongoing debates on the validity of extending duties of care or even rights status to future generations, the argument presented in this report is an interesting one. First, because it does not define the concept of intergenerational solidarity as 'new' or even entirely outside the remit of established institutional thinking. For instance, its consideration of proposals to establish of a new UN High Commissioner for Future Generations to safeguard the long-term interests and needs of global communities takes for granted the limitations of existing structures. Spearheaded initially by civil society organizations, including the Alliance for Future Generations, proposals for a new high commissioner were subsequently taken up by the Preparatory Committee for the United Nations Conference on Sustainable Development who have since fleshed out how a new high commissioner might operate within the existing United Nations system and promote planning for the future in a manner that is in keeping with the core principles of UN Charter. Two high commissioners already exist within the UN system: the United Nations High Commissioner for Refugees (established in 1951) and the United Nations High Commissioner for Human Rights (established in 1993). Elements of both of these offices are thought to offer direct inspiration for the powers and responsibilities of a new High Commissioner for Future Generations, including those relating to international agenda setting, monitoring, early warning and review, encouraging capacity for innovation at the national and subnational levels and reporting to the General Assembly and various higher offices on threats to planetary heritage (see discussion paper submitted to the United Nations Conference on Sustainable Development, note 17). Above all, the understanding is that a new high commissioner would be an advocate of the adjudicated environmental interests of future generations, interacting with Member States, the United Nations and specialized agencies, as well as various stakeholders to ensure that these interests are respected.

As an international agent, the new high commissioner would have a scope of action significantly broader from that exercised at present by national Ombuds Offices. Her/his primary brief would be to foster intergenerational welfare in sustainable development policy internationally and, in that, offer support to agencies operating at regional levels on best practice and new policy measures to limit the collateral impact of present environmental practices on the future. Another proposal explored is that put forward by the Office of the UN Secretary General (Intergenerational Solidarity and the Needs of Future Generations, 2013) is that relating to the appointment of a special envoy to serve as an independent advocate for intergenerational solidarity and promote best practice in future impact assessment. A special envoy would facilitate the engagement of all stakeholders in United Nations negotiations on sustainable development and would report annually to the General Assembly on the progress of its work. A third proposal is the enhancement of 'inter-agency coordination with regard to the needs of future generations'. In this instance, existing offices within the UN system would be invited to promote intergenerational solidarity through the United Nations System Chief Executives Board for Coordination to ensure greater policy coordination across different fields (UN General Assembly, Report of Secretary General, 2013: 17). What is common to all of these proposals is the understanding that new measures to ensure greater intergenerational justice are a natural step forward for the international community and in line with 'already existing' legal stipulations regarding the welfare of 'all members of the human family' (Universal Declaration of Human Rights, 1948). No proposals currently on the table are considered outside the realm of possibility. Rather, each is premised on an endorsement of embedded international institutional arrangements and legal agreements.

Since 1948, states have progressively come to be directed towards the constitutionalization of a range of international legislation incorporating human, social, cultural, economic and political rights. As a consequence, the legislative framework of the modern democratic state today is as much external as it is internal in its focus. One consequence of this is that legal arrangements can be used against states thought to be acting in violation of internationally agreed principles of justice. Perhaps this is one reason why many states are hesitant to enhance the monitoring powers of the UN. Even so, states find themselves today increasingly bound into a three-way system of legitimation on matters of climate justice. To the extent that escalating pollution levels endanger the biosphere of the planet and therefore the future of humanity, all states have a legal duty to prevent further climate destruction (see Oslo Principles on Global Climate Change Obligations, Preamble, March 2015; United Nations World Commission on Environment and Development, 1987). Not only is the 'genuine coordination and cooperation' of all states needed to achieve this goal (UN, 2012, *Back to Our Common Future*: 18), it is legally required (e.g., Principle 3 of the Rio Declaration, 1992; Convention on Biological Diversity, 1992; the Kyoto Protocol, 1997). Second, demands for institutional change continue to grow internally within states amongst citizens concerned for the welfare of future peoples.

In the UK, for example, two-thirds of respondents to a 2013 Ipsos MORI survey considered the needs of 'all future generations' to be more important than those of any one particular generation (Ipsos MORI Generations, 2017). Similarly, the findings of the 2010 Eurobarometer survey suggested publics strongly support 'reforms that will benefit future generations', particularly in Spain (91 percent of respondents), Finland (87 percent), Belgium (83 percent) and the UK (82 percent), even if those reforms mean sacrifices will have to be made in the present. Third, the legitimacy of the state's position on environmental issues is also shaped by more transnationally situated public alliances advocating a firmer legal protection of 'cultural and natural heritage' (e.g., European Youth Forum; Generation RYSE–Earth Guardians, Our Children's Trust, etc.). Generation RYSE (2017), for instance, actively lobbies government agencies at the regional, national and international levels. It describes itself as a transnationally coordinated 'tribe of young activists, artists and musician from across the globe stepping up as leaders and co-creating the future we know is possible'. As representatives of those on the 'front lines of climate change chaos', youth demand a greater recognition of their rights to inherit a world that is equitable, safe and sustainable (see also European Youth Forum, 2017). Institutional changes along the lines just described are considered an essential part of a wider program of action to ensure that such rights are respected. The significance of these actors' demands for reform cannot be underestimated chiefly because of the expectation they create that the excesses of an overheating carbon economy will be democratically contained (e.g., World Future Council, 2016) and replaced with a more expansive ethic of responsibility that delivers on legal and moral duties owed to all members of humanity, including those not yet born.[2] Certainly, if the pressure currently being brought to bear on states to acknowledge the power of law to prevent further climate destruction (e.g., the Vienna Convention for the Protection of the Ozone Layer, the Kyoto Protocol) continues to grow, we may see a situation emerge where both the judiciary and state government resume a more active role in defining future generations as legitimate recipients of present-day protections.

It may be impossible to enable the concurrency of all legally relevant subjects of climate justice at any one time (present *and* future generations), as many are not yet born. However, what we can ensure is that the dependencies of future peoples on the present (for instance, in terms of their identity, physical being and likely ecological, economic and social needs) are protected by institutionalizing a 'reflective paternal' (Rose, 2016: 57) relation between present and future generations (rather than one of domination). In this instance, the actions, laws, constitutions and policy decisions of the present take account of the affects they have on peoples of the future. The aim is to optimize the well-being of present generations in a manner that does not disadvantage those of the future. To be truly effective, a relation of reflective paternalism between generations requires that the right to self-determination be practiced in a more reasoned manner, where distinctions between self and future others are rethought, especially with regard to entitlements to, as well as long-term management of scarce natural resource supplies (Skillington, 2017). Reflective paternalism recognizes the nonsubstitutable relation between present and future generations, the fact that 'what is good for today may not be good for tomorrow'. It draws attention to the multiple implications of communities acting unilaterally or without regard for subsequent generations. It also gives recognition to the fact that while there may be similarities and points of overlap between generations, the social positions, histories and ecological circumstances of each may differ considerably.

Greater credence must, therefore, be given to the democratic entitlements of extended communities (present and future) and to the legal duties of all living sovereign communities to ensure that essential resource supplies are managed sustainably. The expectation would be that present sovereign and international communities show sensitivity to the conditional nature of rights to resources – conditional, that is, on sufficient reserves being preserved with the future in mind. Also, communities must remain conscious, when exploiting their own natural resource wealth, of legal commitments to 'establish conditions under which justice and respect for the obligations arising from treaties and other sources of international law can be maintained, and to promote social progress and better standards of life' for all peoples (see Charter of the United Nations, 1945). Once reflective paternalism comes to be seen as part of the 'normal business' of the legislature (that is, as relevant to working definitions of 'all peoples'), the likelihood is that intergenerational solidarity will begin to incur real legal, political and social consequences. However, at this stage, extending legal consideration to future generations is only beginning to gain institutional ground.

Existing constitutional references to future peoples are an important first step towards building a more comprehensive future ethics. Indeed, they have proven to be an indispensable tool for the advancement of legal cases brought by youth campaigners and their legal representatives against government agencies accused of failing to fulfill duties of care. The realization, however, is that further constitutional changes are required. The legitimacy of certain constitutional rules is sometimes called into question. For instance, states' unilateral right to consume, exploit, sell or lease natural resources on their territories is occasionally challenged if it is felt this right is being used in a manner contrary to the interests of present and future generations or of planetary well-being more generally. Constitutional law is not a 'complete' knowledge order (Buchwalter, 1998: 3). Critical reflection on how constitutional principles of self-determination and freedom, for instance, might be better applied in scenarios of increasing resource scarcity continues to highlight this fact. Ongoing debate on these issues increases the likelihood of future generations eventually being recognized as relevant parties to all justice deliberations and their interests affirmed through a more intergenerationally 'connected' approach to sustainable development. The following chapter considers how such an approach to justice might be developed further in a way that responds more appropriately to emerging ecological challenges.

Notes

1 Edwards and Gillham (2013) present a comprehensive overview of the main conceptual elements of resource mobilization theory (RMT) – a distinct interdisciplinary approach to understanding what 'resources' social movements draw upon to enable effective collective mobilization and achieve desired social outcomes. Amongst those explored are moral resources, cultural resources, human resources, material resources and social-organizational resources (infrastructure, social networks and organizations). Law is not explored as a resource.
2 Democratic presentism, as a quality of modern state democracies, is a bias in favor of the present. It is generally reinforced by the short-term nature of parliamentary election cycles, as well as pressure on governments to produce speedy political outcomes whilst in office in an effort to increase prospects for reelection (Urbinati & Warren, 2008).

References

Ambrusné, Eva Toth (2010) 'The parliamentary commissioner for future generations of Hungary and his impact', *Intergenerational Justice Review*, 10(1): 18–24.
Basic Law for the Federal Republic of Germany – Gesetze im Internet (2017) available at www.gesetze-im-internet.de/englisch_gg/ (accessed: 1 September 2017).
Belgian Constitution – Grondwettelijk Hof (2008) available at www.const-court.be/en/basic_text/belgian_constitution.pdf (accessed: 1 September 2017).
Blokker, Paul (2013) 'Constitutions and democracy in post-national times: A political-sociological approach', *Irish Journal of Sociology*, 20(2): 68–90.
Bruch, Carl et al. (2002) *Constitutional Environmental Law: Giving Force to Fundamental Matters in the EU*, Dordrecht: Kluwer: 261.
Brunkhorst, Hauke (2012) 'Power and the rule of law in Arendt's thought', in Goldoni, Marc & McCorkindale, Chris (eds.) *Hannah Arendt and the Law*, London: Bloomsbury: 215–228.
Buchwalter, Andrew (1998) 'Constitutional paidiea: Remarks on Hegel's philosophy of law', Proceedings of World Philosophical Congress, Boston.
Burke, Edmund (1890) *Reflections on the Revolution in France*, ed. F. G. Selby, London: Palgrave Macmillan.
Castiglione, Dario & Warren, Mark E. (2006) 'Rethinking democratic representation: Eight theoretical issues' (18–19 May) available at http://citeseerx.ist.psu.edu/viewdoc/download?doi=10.1.1.565.9652&rep=rep1&type=pdf (accessed 8 December 2018).
Constitution of Bhutan (2008) available at www.wipo.int/wipolex/en/details.jsp?id=5214 (accessed: 28 June 2018).
Constitution of the Commonwealth of Pennsylvania (1776) available at https://codes.findlaw.com/pa/constitution-of-the-commonwealth-of-pennsylvania/ (accessed: 28 June 2018).

Constitution of France 1958 and its Amendments, available at www.constituteproject.org/constitution/France_2008.pdf?lang=en (accessed: 30 August 2017).

Constitution of Hungary – Constitute Project (2011) available at www.constituteproject.org/constitution/Hungary_2011.pdf (accessed: 1 September 2017).

Constitution of Japan (1946) available at www.solon.org/Constitutions/Japan/English/english-Constitution.html (accessed: 1 September 2017).

Constitution of the Kingdom of Norway (1992) available atwww.constitution.org/cons/norway/dok-bn.html (accessed 8 December 2018).

Constitution of the Plurinational State of Bolivia (2009) available at www.wipo.int/wipolex/en/details.jsp?id=5430 (accessed: 1 September 2017).

Constitution of Poland (1997) available at www.constituteproject.org/constitution/Poland_1997.pdf (accessed: 31 August 2017).

Constitution of the Republic of South Africa (1996) available at www.wipo.int/edocs/lexdocs/laws/en/za/za107en.pdf (accessed: 1 September 2017).

Cornell, Drucilla (1990) 'Time, deconstruction, and the challenge to legal positivism: The call for judicial responsibility', *Yale Journal of Law & Humanities*, 2(2): 267–297.

Deutscher Bundestag (2016) 'Discussion paper by the parliamentary advisory council on sustainable development in preparation for the meeting of the United Nations high level political forum on sustainable development' (9 June) available at www.bundestag.de/en/committees/bodies/sustainability (accessed: 25 February 2017).

Edwards, Bob & Gillham, Patrick F. (2013) 'Resource mobilization theory', in Snow, David A., della Porta, Donatella, Klandermans, Bert & McAdam, Doug (eds.) *The Wiley-Blackwell Encyclopedia of Social and Political Movements*, Oxford: Blackwell: 1–6.

Eurobarometer 74 Autumn 2010 Report – European Commission (2010) available at http://ec.europa.eu/commfrontoffice/publicopinion/archives/eb/eb74/eb74_publ_en.pdf (accessed: 1 September 2017).

European Commission (2017) 'Security union: A Europe that protects' (June) available at https://ec.europa.eu/commission/publications/security-union-europe-protects_en.pdf (accessed: 31 August 2017).

European Youth Forum (2017) available at www.youthforum.org/ (accessed: 20 February 2017).

Foundation for Democracy and Sustainable Development (2015) 'The democratic case for an office for future generations', available at www.fdsd.org/ (accessed: 31 August 2017).

French Republic Constitution (1793) available at https://oll.libertyfund.org/pages/1793-french-republic-constitution-of-1793 (accessed 8 December 2018).

Generation RYSE – Earth Guardians (2017) 'Earth Guardians youth director Xiuhtezcatl Martinez', available at www.earthguardians.org/xiuhtezcatl/ (accessed: 1 September 2017).

Göpel, Maja (2010) 'Guarding our future: How to protect future generations', *Solutions* 1(6): 62–70.

Habermas, Jurgen (2008) 'The constitutionalization of international law and the legitimation problems of a constitution for world society', *Constellations*, 15(4): 444–455.

Hegel, Georg Wilhelm Friedrich (1892) *Lectures on the History of Philosophy*. Vol. 1, trans. E. S Haldane, London: Kegan Paul/Trench Trubner & Co.

Hunger Notes (2015) 'World hunger and poverty facts and statistics' available at www.worldhunger.org/2015-world-hunger-and-poverty-facts-and-statistics/ (accessed: 30 August 2016).

Hunt, Lynn (2007) *Inventing Human Rights*, New York: W. W. Norton & Company.

International Atomic Energy Agency (1997) 'The joint convention on the safety of spent fuel management and on the safety of radioactive waste management' available at www-ns.iaea.org/conventions/waste-jointconvention.asp (accessed: 28 June 2018).

Ipsos MORI (2017) 'Ipsos MORI generations – social research association', available at http://the-sra.org.uk/wp-content/uploads/bobby-duffy.pdf (accessed: 1 September 2017).

Jefferson, Thomas (1776) The Papers of Thomas Jefferson (Volume 1) available at https://jeffersonpapers.princeton.edu/all-volumes (accessed: 8 December 2018).

Jefferson, Thomas (1984) 'Letter to James Madison, 6 September 1789', in Peterson, Merrill D. (ed.) *Thomas Jefferson: Writings*, New York: Library of America: 959–964.

Johnson, Nathan & Asquith, Christina (2010) 'An environmental ombudsman: Sándor Fülöp, Hungary's parliamentary commissioner for future generations', *Solutions*, 1(6) (November) available at www.thesolutionsjournal.com/article/an-environmental-ombudsman-sandor- (accessed: 17 February 2017).

Kant, Immanuel (1970[1784]) *The Idea of a Universal History with a Cosmopolitan Purpose*, *Kant's Political Writings*, ed. Hans Reiss, trans. H. B. Nisbet, Cambridge: Cambridge University Press.

Kyoto Protocol – United Nations Framework Convention on Climate Change (1997) available at http://unfccc.int/kyoto_protocol/items/2830.php (accessed: 1 September 2017).

Lenoble, Jacques & Maesschalck, Marc (2016) *Democracy, Law and Governance*, Oxon: Routledge.

Luhmann, Niklas (1987) 'Closure and openness: On reality in the world of law', in Teubner, Gunther (ed.) *Autopoietic Law: A New Approach to Law and Society*, Florence: De Gruyter/European University Institute: 12–35.

Luhmann, Niklas (1997) 'The control of intransparency', *Systems Research and Behavioral Science*, 14(6): 359–371.

Marks, Susan (2000) *The Riddle of All Constitutions: International Law, Democracy and the Critique of Ideology*, Oxford: Oxford University Press.

Marx, Karl (1852) '*The Eighteenth Brumaire of Louis Bonaparte*' available at www.marxists.org/archive/marx/works/1852/18th-brumaire/ (accessed: 16 February 2017).

McCarthy, Thomas (1994) 'Philosophy and critical theory: A reprise', in Hoy, David C. & McCarthy, Thomas (eds.) *Critical Theory*. Oxford: Wiley-Blackwell: 5–86.

National Assembly for Wales (2015) 'The wellbeing of future generations (Wales) bill' available at www.legislation.gov.uk/anaw/2015/2/contents/enacted (accessed: 2 September 2017).

Nesbit, Martin & Illés, Andrea (2015) 'Establishing an EU guardian for future generations', Report and Recommendations for the World Future Council, London, Institute for European Environmental Policy.

Oslo District Court Proceedings (2018) 'Case No. 16–166674TVI-OTIR/06' available at https://secured-static.greenpeace.org/norway/Global/norway/Arktis/Dokumenter/2018/Judgement%20-%204.%20jan%202017%20-%20Oslo%20District%20Court%20stamped%20version.pdf?_ga=2.94554334.338564569.1530190208-524273936.1530190208 (accessed: 28 June 2018).

Parliamentary Advisory Council on Sustainable Development (2017) 'Committees' available at www.bundestag.de/en/committees/bodies/sustainability (accessed: 1 September 2017).

Report of the World Commission on Environment and Development: Our Common Future (1987) available at www.un-documents.net/our-common-future.pdf (accessed: 1 September 2017).

Rio Declaration on Environment and Development – The United Nations Conference on Environment and The Development (1992) available at www.un.org/documents/ga/conf151/aconf15126-1annex1.htm (accessed: 1 September 2017).

Rose, Michael (2016) 'Constitutions, democratic self-determination and the institutional empowerment of future generations: Mitigating an aporia', *Intergenerational Justice Review*, 2: 56–71.

Save the Arctic (2017) 'The climate lawsuit against the Norwegian government' available at www.savethearctic.org/en/peoplevsarcticoil/background-documents/ (accessed: 28 June 2018).

Skillington, Tracey (2017) *Climate Justice & Human Rights*, Basingstoke: Palgrave Macmillan.

Teubner, Gunther (1992) 'The two faces of Janus: Rethinking legal pluralism', *Cardozo Law Review*, 13: 1443–1462.

Thompson, Dennis (2005) 'Democracy in time: Popular sovereignty and temporal representation', *Constellations*, 12: 241–258.

United Nations (1972) 'Declaration of the United Nations conference on the human environment)', UN Doc. A/Conf.48/14/Rev.1) available at www.un-documents.net/unchedec.htm (accessed: 1 September 2017).

United Nations Charter (1945) available at www.un.org/en/charter-united-nations/ (accessed: 1 September 2017).

United Nations Conference on Sustainable Development (2012) 'The future we want' (Rio de Janeiro, Brazil) available at https://rio20.un.org/sites/rio20.un.org/files/a-conf.216l-1_english.pdf.pdf (accessed: 30 August 2016).

United Nations Convention on Biological Diversity (1992) available at www.un.org/en/events/biodiversityday/convention.shtml (accessed: 1 September 2017).

United Nations Convention to Combat Desertification (1994) available at www.agrhymet.ne/portailCC/images/pdf/FichePédag5_en.pdf (accessed: 28 June 2018).

United Nations Convention on the Law of the Sea (1982) available at www.un.org/depts/los/convention_agreements/texts/unclos/unclos_e.pdf (accessed: 28 June 2018).

United Nations Department of Economic and Social Affairs (2012) 'Back to our common future: Sustainable development in the 21st century (SD21) project' available at https://sustainabledevelopment.un.org/content/documents/UN-DESA_Back_Common_Future_En.pdf (accessed: 1 September 2017).

United Nations Development Group (2015) 'The world we want: Dialogues on the implementation of the post-2015 development agenda' available at www.un.org/millenniumgoals/pdf/UNDG%202nd%20dialogues.pdf (accessed: 28 June 2018).

United Nations Framework Convention on Climate Change (1992) available at https://unfccc.int/resource/docs/convkp/conveng.pdf (accessed: 1 July 2018).

United Nations Framework Convention on Climate Change. The Paris Agreement (2015) available at: https://unfccc.int/sites/default/files/english_paris_agreement.pdf (accessed: 8 December 2018).

United Nations General Assembly (2009) 'Report of the office of the United Nations High Commissioner for Human Rights on the relationship between climate change and human rights', A/HRC/10/61 (15 January) available at www.ohchr.org/Documents/Press/AnalyticalStudy.pdf (accessed: 1 July 2018).

United Nations General Assembly (2013) 'Intergenerational solidarity and the needs of future generations: Report of the Secretary-General' (15 August) available at https://sus

tainabledevelopment.un.org/content/documents/2006future.pdf (accessed: 27 December 2017).

United Nations Open Working Group on Sustainable Development Goals (2014) available at https://sustainabledevelopment.un.org/owg.html (accessed: 1 September 2017).

United Nations Universal Declaration of Human Rights (1948) available at www.un.org/en/universal-declaration-human-rights/ (accessed: 28 June 2018).

United Nations World Commission on Environment and Development: Our Common Future (1987) available at www.un-documents.net/our-common-future.pdf (accessed: 28 June 2018).

United Nations World Meteorological Organization (2016) 'Climate breaks multiple records in 2016, with global impacts', available at https://public.wmo.int/en/media/press-release/climate-breaks-multiple-records-2016-global-impacts (accessed: 30 October 2017).

United Nations World Summit for Social Development (1995) Copenhagen, 6–12 March, Chapter One, Resolution 1, annex 1, para. 26(b).

Urbinati, Nadia & Warren, Mark E. (2008) 'The concept of representation in contemporary democratic theory', *Annual Review of Political Science*, 11: 387–412.

Williams, David (2004) *Condorcet and Modernity*, Cambridge: Cambridge University press.

Wolfrum, Rudiger & Grote, Rainer (Eds.) (2011) *Constitutions of the Countries of the World*, Dobbs Ferry, NY: Oceana.

World Future Council (2011) 'Ombudspersons for future generations – Stakeholder Forum' available at https://sf.stakeholderforum.org/fileadmin/files/SDG%204%20Ombudspersons%20for%20Future%20Generations%20Thinkpiece.pdf (accessed 8 December 2018).

World Future Council (2013) 'Guarding our future: How to include future generations in policy making' available at www.futurejustice.org/wp-content/uploads/.../brochure_guarding_en_final_links2.pdf (accessed 8 December 2018).

World Future Council (2016) 'Annual report 2016', available at https://www.worldfuturecouncil.org/annual-report-2016/ (accessed: 8 December 2018).

Chapter 6

A deeper framework of intergenerational justice

Introduction

The discovery of the Anthropocene has prompted a fundamental reconsideration of capitalist civilization as the culmination of positive forces of evolutionary change. In particular, its association with a monumental collapse of geological history into human history cycles (Chakrabarty, 2011) and the way this development necessitates an elementary change in knowledge perspective on the relationship between agency and time. There are several reasons why this change in perspective is essential. Most notably, the realization that the effects of human destructive practices reverberate across multiple time frames and implicate many generations of victims. We are forced to consider how the scope of ethical consideration can be extended to accommodate a future deeply affected by the pollution practices of previous generations. As indisputable scientific facts (see NASA, 2017; IPCC, 2014; Raupach & Canadell, 2010), these developments provoke serious questions about where the boundaries of the just society ought to reside?[1] Increasingly, the future is conceptualized as a present one (Beck, 2015), as scenarios of melting polar ice and record heat waves offer an early glimpse at unfolding ecological dystopias. Yet consistently these insights do not inspire the international community to take the necessary steps to reduce GHG emissions (global emissions of carbon dioxide have risen more than 60 percent since international climate change negotiations began in 1992; see the 2017 Atlas of Sustainable Development Goals: A New Visual Guide to Data and Development, World Bank). In many ways, we remain captive to the demands and distractions of the present. A bias (known as 'presentism') is reproduced across all knowledge systems today, including law, economics, politics but also the social sciences (with an emphasis on the living subject). Just as Beck (2006) identified a need to challenge 'methodological nationalism' with a more cosmopolitan sociological perspective on the social world, this book has explored a range of issues that draw attention to the need for new forms of conceptual and empirical analysis of relations between generations across time and, with that, a need to reflect critically on tendencies towards methodological presentism (e.g., limiting justice perspectives to the present time frame).[2] It has looked at how these concerns inform a critical diagnostics of society as marked by deep inequalities between generations of the Anthropocene and, further, how youth respond to this situation.

The absence of a macro perspective on relations of justice across generations

One of the great virtues of democratic societies is the power they accord to publics to hold existing government representatives accountable for their actions and decisions. However, this virtue can become a vice, as Young (2011: 137) observes, when this power is used to prioritize the interests and needs of some over those of others (e.g., future generations). The power of accountability can inadvertently diminish prospects for developing a more macro perspective on relations of justice across a broader time range and allow short-term concerns to dominate the issue agenda. Intergenerationally, this can give rise to inequalities amongst peoples of different ages living within the same society. The contemporary 11-year-old's experiences of climate change over the course of her lifetime will, in all likelihood, be entirely different from those of a 60-year-old today, the majority of whose life is in a past largely undisturbed by knowledge of the Anthropocene and who has benefitted from many years of carbon pollution. Major differences emerge between generations in terms of experiences of crises, deprivation and inequality, giving rise to markedly different sentiments of belonging and identification with a world in serious danger of collapse (Beck, 2015: 85). The accumulation of many years of GHG pollution produces a progressively more unstable planetary system for the future, forcing newer generations to adapt to declining living conditions created largely by their parents and grandparents' generations. In this situation, youth and generations to come suffer adversities that are inherited rather than self-imposed.[3] The profoundly unjust dimension of this situation does not go unnoticed by those most affected by it. Youth, especially, formulate present ecological scenarios as the product of relations of domination and an unfair limiting of justice standards to those who exert power in the present. Even in the legal sphere, emphasis is placed on the physical presence of the subject, as well as direct evidence of wrongdoing.[4] 'Democratic presentism' (Thompson, 2010: 2) further exacerbates these tendencies. With its emphasis on short-term policy goals and electoral cycles, it exonerates all in their lack of attentiveness to the dire long-term implications of high-carbon energy policies and natural resource depletions, as well as the involuntarily inclusion of billions in cycles of deepening humanitarian and ecological destruction.

Barriers in the way of a social connection approach to justice

The discovery of the Anthropocene offers unprecedented evidence of capitalism's destructive interference with the Earth's ecosystems (Chernilo, 2017: 44). However, it also provides stark evidence of how this destruction compounds relations of inequality between global regions (OECD, 2004) and generations (UN, 2013).[5] A knowledge perspective that responds to such insights by continuing to focus only on relations between peoples in the present, whilst ignoring deepening inequalities emerging between present and future communities of humanity, is unjust. A longer-range cosmopolitan perspective on the plural temporalities of climate destruction extends understandings of ecological agency, responsibility and rights eligibility outward. In many ways, it is more akin to a social connection approach (Young, 2011), where deteriorating climate conditions, in implicating multiple victims, agents and sources of harm across time and place, necessitate forms of democratic organization that can accommodate the needs of many generations. There are ample reasons why such an approach must be pursued as a matter of urgency, not least the mounting evidence of harm, yet several barriers currently stand in the way of its realization.

Never has society been so knowledgeable about the nature of ecological threat (giving rise to a new level of 'catastrophism in social thinking', according to Urry (2016: 34), yet consistently it has failed to act to change the course of planetary destruction. Perhaps one reason for this is the ongoing tendency to formulate climate change as an exogenous force that presents itself as a series of shocks to an otherwise independent society. Still, the desire is to reestablish states of equilibrium between society and nature as though they are separate entities. Deepening contradictions, however, problematize the plausibility of an equilibrium model as climate change arises in the first instance because of capitalism's insatiable appetite for natural resources and its drive to promote 'consumptions of excess' (Urry, 2010: 192). As global supplies of essentials decline, the moral imperative to 'do unto future generations as you would have past generations do unto you' (Rawls, 1993: 274, 2001: 159–160) is replaced with the mantra 'each to their own!'. Duties to share nonrenewable resources (e.g., freshwater, natural gas, oil, phosphorus, earth minerals) with future others are losing their institutional potency as the race for resources intensifies internationally. Major inconsistencies emerge between those who exert rights over limited resource reserves and those denied a portion of the same. If democratic justice is the accomplishment of a system of social cooperation across time, its absence at present can only be interpreted as the achievement of a system of noncooperation and exclusion. This view certainly resonates with those asserting their voice on these matters. As those whose welfare is threatened by the actions of the powerful and whose rights to inherit a safe world are systematically ignored, youth define their status as that of 'the oppressed'. Building on sentiments of injury and moral disrespect (Honneth, 2007), mobilized youth question working assumptions as to who is legitimately entitled to make claims to justice in this era of deepening ecological, social and economic problems (individual citizens, bounded political communities, wider transnational ones or generations)? In doing so, they bring to the fore fundamental questions regarding status inequalities, misrecognition and unfair exclusions. Not only do they question the validity of existing democratic arrangements (e.g., the lack of transparency surrounding energy policy or pollution control), they also consider whether established norms can be reinvigorated in response to a growing range of injured subjects. In raising these concerns, youth reestablish the metapolitical value of modernity's ideals of democracy (Held, 1995) to Anthropocene futures.

Throughout this book, emphasis has been placed on how the project of transformative action is envisaged socially, politically and legally by a range of social actors who actively seek to bring an additional layer of critical interpretation to bear on climate change issues as issues of justice, rights and responsibilities (e.g., responsibility for grave levels of resource destruction, accountability for rising CO_2 emissions levels and the future as a relevant context of justice). The focus has been on how these actors envisage the transformative potentials of the existing basic structure of justice and put forward a series proposals as to how a more comprehensive program of intergenerational solidarity and institutional responsibility for environmental protection can be realized by the same. One of the more interesting aspects of mobilization on these issues, therefore, is the extent to which campaigners presuppose that the facts of ongoing deteriorations in climate conditions and the validity of existing democratic norms can be reconciled through a type of 'transcendence from within' (Habermas, 1992: 14) the existing societal order – that is, through a type of creative process of reflection on and critical reinterpretation of justice ideals in response to a range of climate change concerns.

As carriers of a new vision of democratic justice, youth define themselves as contemporary agents of democracy, creators of new what-if ethical horizons for climate futures. The way in which these actors define their agenda, both collectively and individually, as an act of resistance to domination and rights violations is of immense sociological interest. Youth use the language and institutional tools of democracy to reinforce a definition of themselves as subjects of liberty and creativity (Touraine, 1996: 293–294). They assert the right to define the social experiences of their generation and of those to come, of risk endangerment, deprivation and exclusion, as authentic (see, also, UNICEF, 2010). In this sense, democracy remains the essential basis for the rise and flourishing of social actors, those who stress the right to be active agents in the shaping of Anthropocene futures. More than just 'rational participation in social life', democracy is also 'the recognition of personal subjects and of the diversity of their attempts to reconcile instrumental reason with a cultural identity, personal and collective, which implies the greatest liberty for all' (Touraine, 1997: 126). 'Personal subjects' of climate change formulate their experiences of pollution and exclusion as exemplars of wrongdoing, using those experiences to subjectivize human and political rights norms (Touraine, 1997: 186) and, in doing so, also build capacity in terms of the formulation of a cogent critique of those action sequences and power structures that threaten their interests.

The 'imagination of justice' that arises from this process always remains rooted in the social historical world but in a way that posits 'new determinations' of old principles (of rights, responsibilities, duties, etc.) (Castoriadis, 1987: 264–265).

The contribution of these actors to new formulations of justice cannot be underestimated. In the legal domain, future justice and youth coalitions, especially, in cooperation with various transnationally relevant legal experts (e.g., Client Earth, Our Children's Trust, Law Response Team), have sought to test legal commitments to, for instance, rights to a safe and healthy future, survival and development ('endogenized' [Sassen, 2003] within specific national/regional settings) against the burden of proof of deepening ecological destruction and a growing range of expert knowledge claims. The increasing regularity of this comparative exercise stimulates the democratic imperative of law, especially the need to reconcile contradictions emerging between constitutionally grounded rights to the appropriation and exploitation of natural resources, on the one hand, and rights to development, health, a safe environment on the other. The realization is that the constitutionality of present and future generations interests (e.g., to be self-determining) must be protected. Legal cases are an important reminder to states of constitutional requirements to remain open to newer demands of justice, including those emerging from below amongst aggrieved youth who feel locked out of decision making on issues of crucial importance to their future, issues that may lead to steering failure if greater efforts are not made now to control them. Justice, youth campaigners claim, does not have a future without an intergenerationally relevant contract that constitutionally guarantees that best efforts will be made to conserve essential resources for the years ahead.

Pressure to initiate a more responsible regulation of intergenerational conduct, in turn, focuses attention on the question of what development practices ought to be defined as 'acceptable' and in keeping with international human rights regulations. For many, the best way to determine 'acceptable development practices' is to first decipher what is unacceptable. Actors such as Greenpeace (2016), World Future Council (2013) and Friends of the Earth Europe (2009), for instance, have all earmarked certain practices, including Arctic drilling, razing the rainforests, and bottom trawling of oceans as unacceptable. As practices that will 'cause serious widespread and long term harm to the health, safety or survival of future generations', these are said to be equivalent to 'crimes against future generations' (e.g., see World Future Council, 2013). These actors consciously extend the definitional boundaries of legal interpretations of 'crimes' to include newer varieties. Mehta and Merz (2015: 3) define 'ecocide' as 'significant damage to or destruction of an ecosystem to such an extent that peaceful enjoyment of a part of the planet will be substantially diminished'. As early as 1973, following the 1972 UN Conference on the Environment, a draft ecocide convention was published by Richard A. Falk, calling for ecocide to be added as the fifth crime against peace to the Rome Statute of the International Criminal Court (ICC) in The Hague. Article 8, 2, iv of the Rome Statute does refer to 'widespread, long-term and severe damage to the natural environment' as a crime during wartime but, peculiarly, not during times of peace (Greppi, 1999; Merz, Cabanes & Gaillard, 2014; Mehta & Merz, 2015: 7). Originally proposed by U.S. biologist Arthur Galston, ecocide was reintroduced by Swedish Prime Minister Olof Palme during his opening speech at the 1972 UN Stockholm Conference on the Human Environment. It would surface again in 2008 when Scottish lawyer Polly Higgins, seeking a more solid legal framework for the implementation of long-term environmental protections, revived the term. Higgins proposed that the International Law Commission modify the Rome Statute to include the crime of ecocide. In 2013, the World Future Council, together with Sébastien Jodoin, Lead Counsel with the Centre for International Sustainable Development Law (CISDL), proposed a series of new policy measures to address 'crimes against future generations' and create a stronger climate of expectation that international obligations to protect natural resource heritage will be respected:

> The term 'humanity' in crimes against humanity indicates that this crime concerns offences which are of concern to all of humanity, and that the gravity is such that when they are committed, all of humanity is injured and aggrieved. Crimes against future generations are similar, and arise where there is a connection, in terms of knowledge and causation, between the underlying offence and damage in the long-term.
>
> (World Future Council, 2013)

The European Citizens Initiative End Ecocide was launched in 2013 (over 185,000 citizens' signatures were gathered supporting a law of ecocide prevention) calling on the European Parliament to introduce new legal provisions to prohibit, prevent and preempt the crime of ecocide, 'a crime for which those in positions of superior responsibility can be held accountable'. At the Paris UN Climate Summit (COP 21) in 2014, End Ecocide submitted its ecocide amendment to then UN Secretary General Ban Ki-Moon. The amendment focuses on the protection of the atmosphere, the oceans, seas beyond territorial waters and the Arctic, all of which are recognized as *res nullius* in law, that is, as resources that belong to no one and therefore should not be sites for commercial exploitation and state rivalry, yet are currently threatened by both.

If the crime of ecocide were to be officially recognized, how would a judgment be made regarding 'unacceptable' levels of ecological harm (i.e., constituting a crime against present and future humanity)? Second, how will excessive violations of the rights of future generations be addressed (what kinds of penalties will be imposed)? In drawing attention to these issues, these actors highlight the importance of addressing the procedural as much as the moral practical dimensions of this approach to intergenerational justice. The fact that the welfare of future generations is already referenced across a range of legal instruments (including state constitutions, declarations, conventions) is not, arguably, in itself sufficient to ensure that a necessary degree of justice will prevail (e.g., the establishment of a new court, as well as a new High Commissioner for Future Generations, the implementation of independent future impact assessments across all policy areas or the immediate revision of energy targets downward in recognition of the needs of future generations). For defenders of the ecocide campaign, intergenerational justice must be a matter of assignable and enforceable rights.

Addressing regulatory dysfunctionality

The 'deep history' of global climate destruction, encompassing the past, present and future, requires a more comprehensive framework of justice deliberation and representation, one that approaches the distribution of ecological harms in terms of the facts of inequality and unfair advantage encouraging, in the process, a further elaboration of the interpretation of rights. This means addressing current problems with procedural arrangements, in particular, regulatory agencies' poor record of environmental protection and consistent failure to address the underlining, sociogenic roots of ongoing climate destruction. An example is the continuing political support for and minimal regulatory control of the activities of the fossil fuel industries. Reconciling further extreme fossil fuel extraction and consumption with the changes needed to avoid a 2°C rise in global average temperatures makes little practical sense. By any stretch of the imagination, a regulatory regime that only minimally registers the dangers inherent in this development pathway and the threat it poses to biodiversity across a range of scales is dysfunctional.

Dysfunctionality in this instance does not arise from a lack of awareness of the long-term effects of short-term policy thinking but from a type of regulatory control that seeks to protect the interests of the fossil fuel industries by imposing tariffs and quotas on low-carbon alternatives (e.g., the decision of the Trump administration in 2017 to scrap incentives to switch to renewable energy systems in the interests of protecting U.S. fossil fuel industries) and by ignoring the dangers posed by hydraulic fracking in terms of the risk of contamination of drinking water wells, methane pollution and increased exposure to toxic chemicals. Regulatory dysfunctionality arises from the non-representative nature of present decision-making arrangements and from governments' lack of attention to the foundational core of publics' experiences of and increasing vulnerability to scenarios of severe natural resource scarcity, scorched lands, poor crop yields, plummeting water supplies and so forth. If the type of transformative spirit that embodied past democratic struggles seems wholly absent today, it is because of these problems with the regulatory control of environmental destruction, the failure of states to deliver on obligations to protect and ensure equality of participation for all. For aggrieved citizens, however, current institutional failings are seen not as evidence of the impossibility of justice but rather as the incomplete realization of democratic potentials (Bohman, 2012). A chief focus of critique is the lack of reciprocal-general justification for further extreme energy projects or unconvincing assurances from international agencies (e.g., the International Energy Agency, 2017) as to the abundance of remaining supplies of 'cheap fossil fuels', the necessity of further deep-sea gas and oil exploration and the economic importance of the 'gas market revolution'. Campaigners point to the lack of dialogue between this discourse and the documented links between fossil fuel pollution and rates of cardiovascular, cancer and respiratory diseases in humans, as well as reproductive failure in other species life (e.g., see Lancet Commission Report, 2017; Union of Concerned Scientists, 2016; World Health Organization, 2014).[6]

The ongoing promotion of a fossil fuel agenda in the face of increasing evidence of the dangers of doing so is not grounded sufficiently in procedures of democratic justification to qualify as fair to all parties concerned (see, for example, EarthJustice, 2017). Scientists advise, on the basis of cumulative research, that one-third of remaining oil, half of gas and over 80 percent of current coal reserves stay in the ground to limit global warming to 2°C and to avoid further 'thermal expansion' (e.g., a greater incidence of melting ice sheets and significant sea level rises; see, for example, McGlade & Ekins, 2015). Campaigners look to law to force nonreciprocal and nonrepresentative justifications for the further exploitation of territorial fossil fuels to be made more explicit and accountable to publics. In that, law offers concerned citizens a vital means of realizing constitutional rights to challenge government actions believed to be harmful to life and liberty, as well as secure a degree of testimonial justice on environmental regulation, currently denied to them through the closed-door decision-making policy of many states.

While legal action is an essential first stage in challenging dysfunctional regulatory arrangements, by itself, it is not sufficient to initiate a more structurally grounded regime of climate justice. To have an enduring influence, justice requires a more advanced framework of institutional change, with regulatory procedures that are genuinely democratic in organization and publicly legitimized. Ultimately, it is publics who must insist that such institutional changes occur just as the expectation is that it is publics who will nurture democratic cultures into the future and ensure that the democratic structures of society remain reflexive and self-correcting. Responsibility for initiating a deeper climate justice framework, therefore is a collective one, requiring the cooperation of many publics. Civil society has a crucial role to play in bringing such cooperation about and ensuring government at all levels remains open to a range of new concerns and problem areas, as well as proposals for addressing climate destruction. Until such time as they do, concerned citizens will be forced to continue to bring their grievances to court and, in that, communicate a strong public message on the failure of state agencies to perform the most basic of functions – to protect the peoples of their sovereign communities and uphold fundamental rights. What publics demand now are forms of procedural justice that extend public influence over policy on future energy scenarios and essential resource management, as these issues become increasingly life determining. The aim must be to ensure that a strict compliance with environmental regulations is maintained and legally grounded standards of intergenerational justice respected.

Representing the interests of youth and citizens-to-be

As future generations cannot at this point in time be party to such procedures, there is a need for some form of institutional representation of their interests. One proposal is the nomination of trustees who would act on their behalf. As Thompson (2010: 27) highlights, trusteeship prescribes the equivalent of bifocals to address current defects in democratic vision, where the capacity to see different kinds of collective action problems from a longer-range perspective is enhanced and new environmental challenges approached with both current and future citizens in mind. Ultimately, the aim of a trustee model of intergenerational justice is to ensure that both present and future citizens have sufficient opportunities to be self-determining by safeguarding reserves of essential resources. This model would, at least, begin to tackle the problem of short-term policy thinking or single-generation justice deliberation by giving greater currency to the idea of an intergenerational contract, one that constitutionally guarantees best efforts will be made to conserve essential resources. Without such a contract, there is a real danger many state communities will lose their capacity to be ecologically resilient (e.g., sinking states). At present, the territorial state may be the unchallenged standard for collective political organization, but for this arrangement to continue to be universally viable into the future, the resource base of all self-determining state communities must be protected and the needs of 'disappearing states' accommodated. No coordinated plan of rescue exists at present to accommodate the territorial needs of ecologically displaced state communities (see Skillington, 2016: 177–206). If the territorial state is to continue to have global relevance in a world where habitable territories are increasingly scarce (assuming global warming intensifies further), then universally shared notions of justice (including human rights) will have to be imaginatively rethought (Fine, 2009: 20) and international cooperation defined in more radically cosmopolitan terms (e.g., accommodating newly displaced communities). If some states are losing their territories and essential resource reserves through no fault of their own (as a consequence of *globally determined* climate changes), how can such communities (e.g., small island states) implement trustee arrangements effectively with future generations firmly in mind? Historically, accumulating pollution is gradually overwhelming these communities' capacities to adapt. The understanding must be that trusteeship, whilst inclusive of the needs of future citizens of specific states, also extends at some level to all peoples, a viewpoint particularly relevant to campaigns for atmospheric justice (e.g., Atmospheric Trust Litigation [ATL]).

Campaigners draw attention to local, nationally and transnationally relevant histories of rescue, resistance and democratic transformation as a stimulus to action. The imagination of historical justice is extended to an imagining of current worlds under threat in the hope of triggering a greater desire for cooperative action amongst all those broadly committed to a project of ecological rescue and democratic rejuvenation. Responsibility for the coordination of such efforts, ultimately, is a shared one deriving from our belonging together with others in a world where the safety of all is threatened (Beck, 2016: 165). Assuming that states are willing to adopt such an approach and recognize the importance of global cooperation on ecological issues in a cosmopolitan spirit, it is perhaps reasonable to assume that the application of a trustee model is feasible. However, such an approach would have to encompass a more comprehensive, transnationally coordinated model of resource management, one that recognizes how no one level of governance in isolation can address current global climate adversities or secure the long-term availability of essential resources without the cooperation and assistance of others. Arguably, this is the only way of insuring against catastrophic futures and securing a deeper intergenerational justice framework. That said, it is at the state level that the constitutionality of present and future generations' resources interests are most powerfully articulated at present. State constitutional law offers the strongest legal clarification of the importance of long-term environmental obligations and in this sense is the most important legal mechanism for promoting a trustee model of resource justice in the first instance. Article 4 of the Convention Concerning the Protection of the World Cultural and Natural Heritage (1972) reiterates this point when it clarifies how the duty to ensure 'the identification, protection, conservation, presentation and transmission to future generations of the cultural and natural heritage' belongs primarily to the state. Furthermore, the understanding is that states will perform this duty to the best of their abilities and 'with any international assistance and co-operation' that can be obtained.

The state's role in implementing such legislation must continue to be elaborated, as new insights on climate risk and security urgencies arise. However, states also require further support from the international community at large. The adoption of Agenda 2030, as well as a series of new climate goals following the COP 21 negotiations in Paris in December 2015 point to the importance of a more coordinated strategic plan for long-term resource management.[7] Proposals to establish a new UN High Commissioner for Future Generations (UN, 2013) are in keeping with this renewed emphasis on the need for an internationally coordinated response from 'cosmopolitan climate risk communities' (Beck, 2016), that is, communities bound together by shared legislation, ecological risks, political decision making and forms of civic solidarity beyond the borders of individual states. As Beck (2016: 165) explains, the cooperative projects of these communities are always in process, moving forward, evolving in response to new contingencies as they arise.

For instance, discussion on proposals to establish a new UN High Commissioner for Future Generations debate whether such a role should be exclusive to just one UN office or distributed across several, each addressing different areas of specialization. Whatever its final configuration, the aim of such an office will be the same – to ensure that the core principles of the UN Charter are seen as encompassing past, present and future generations of humanity (see UN General Assembly, July 27, 2012, 66/288 'The Future We Want'). The legal rationale for a new High Commissioner for Future Generations' is said to be grounded in existing international legal commitments to all members of the 'human family' across time (see, for example, the Preamble of the International Covenant on Economic, Social and Cultural Rights, which refers to 'the equal and inalienable rights of all members of the human family') and to a cosmopolitan project of perpetual peace.

Ratified conventions, including the Convention on the Rights of the Child, require signatories to periodically report to the UN on how they are advancing rights across all policy areas. Reports are subsequently reviewed, and the UN responds with recommendations, advising where further improvements can be made. There is no reason to assume that a similar approach could not be applied in relation to environmental law. Posterity impact assessments could easily be added to established procedures. An example is the type of environmental impact assessments currently in operation in the EU, the United States and the UK. Governments in this instance would be asked to issue statements on how prevailing policies might negatively affect the environmental and social resilience of future generations. At the EU level, calls for a further extension of policies and values supportive of a deeper justice framework (e.g., the precautionary principle, environmental impact assessments, sustainability performance indicators) are gaining ground. Campaigners draw attention to the need to substitute short-term thinking and its drivers (self-interest, electoral dependence, immediate performance indicators, etc.) with more future-oriented goals (see World Future Council, 2015) by institutionalizing a range of long-term impact assessment procedures across all institutions. At the regional level, also, government agencies are under increasing pressure to assist public bodies in the achievement of sustainable development goals (e.g., in Wales, the establishment of the Welsh Commissioner for Future Generations).

Internationally, moves to substitute short-term policy thinking with more long-term approaches are still very much at the proposal stage. Ongoing support for the carbon economy remains a primary concern of campaigners operating at this level. The latter point to the fact that the right to a healthy, safe environment is a collective as much as individual right, one that meets what Raz (1986: 208) describes as the three conditions necessary to count as a valid rights claim: First, an aspect of the interests of future generations (e.g., the need for environmental conditions supportive of health and human flourishing) justifies holding present generations to a duty to protect such interests. Second, the interests in question are interests in public goods shared by all, and having a right to such goods serves the collective interests of future generations. Third, the interests of no single member of future generations in this public good (e.g., a clean atmosphere) is sufficient by itself to justify holding present generations accountable for duties of care and legal rights fulfillment, but the interests of all members together as a collectivity are The extension of duties of care and rights recognition to future generations (as a collectivity) is guided by an attitude of 'reflective paternalism' (Rose, 2016: 57), where present and future peoples are seen as bound together not only by a shared biological identity but also by shared ecological resources and a common cosmopolitan project as well. The latter 'brings together institution[s] and outlook, judgment and understanding' (Fine, 2006: 64) in confronting the major challenges of our age. It creatively reimagines the boundaries of collective belonging to world risk society (Beck & Levy, 2013) by extending relevant criteria outward in recognition of growing interdependencies across space and time.

Ultimately, the political, social and cultural salience of this more cosmopolitan approach to intergenerational justice depends on our capacities to learn from transformations taking place in the experiential spaces of everyday life (e.g., deteriorations in environmental conditions) and on our willingness to act to change catastrophic futures by protecting shared aspects of our planetary existence from forces destined to destroy it. A deeper vision of democratic justice looks firmly beyond the settled convictions of current short-term policy reasoning to a consideration of how sustainable futures can be actively realized. If current rates of resource depletion and global atmospheric pollution tell us anything, it is that such a framework of climate justice is not being applied at present. The primary objective, therefore, must be to change this state of affairs. The achievement of necessary goals in this regard is best secured on the basis of a variety of configurations of solidarity (local, national and transnational) and of open procedures of democratic deliberation, procedures that genuinely reflect the imperatives of public reason and that frame cosmopolitan principles of freedom, responsibility and right as intergenerationally relevant. As associates of local, national and transnational democratic collectivities and as members of communities commonly threatened with ecological disaster, we all possess a legitimate claim to the decision-making processes of such collectivities. Our associative relationship with one another in this regard forms the strongest ethical basis for action to protect the capacities of all to remain resilient in the face of grave ecological problems in the years ahead and to nurture the abilities of all peoples to realize more sustainable futures.

Notes

1 In its Fifth Assessment Report (2014), the Intergovernmental Panel on Climate Change, a group of 1,300 independent scientific experts from around the world, concluded that warming of the global climate system is 'unequivocal' and that human influence is producing this outcome is 'clear' (ibid.: 2).
2 Beck and Sznaider (2006b) explain how methodological nationalism reflects a tendency amongst social scientists to consider the nation state the most relevant unit of sociological analysis and assume dualities of global and national or them-and-us still apply. The fact that such dualities have dissolved under the weight of cosmopolitanizing forces of change means new forms of conceptual and empirical analysis are required. One might say the cosmopolitan condition today requires a similar questioning of such dualities as they apply to the present and future.
3 The imbalance between current and future generations has clear environmental but also financial implications, with the costs of reducing emissions and adapting to climate change set to impose a heavy fiscal burden on governments of the future.
4 A liability approach to environmental issues (still the most dominant in environmental law), for example, places too heavy an emphasis on the physical presence of the victim and direct evidence of wrongdoing.
5 According to a 2004 report by the OECD (Poverty and Climate Change), 96 percent of disaster-related deaths in recent decades have occurred in the developing world, confirming the predictions of the IPCC's Third Assessment Report that developing regions will suffer the negative impact of climate change most. Especially vulnerable are tropical and subtropical areas where the loss of life and livelihood (depletion of fishing grounds, crop yields due to drought or storm damage) from climate changes is most pronounced. The human, institutional and financial capacity of these regions to anticipate and respond to the direct and indirect effects of climate change remains low.
6 Soaring temperatures are also expanding the geographic range of vector-borne diseases, including malaria and meningococcal meningitis, which tend to be concentrated

amongst children, whilst drought and flooding are increasing the incidence of diarrheal diseases (see UNICEF, 2017).

7 The United Nations, for instance, highlights how

> the future of humanity and our planet lies in our hands. It lies also in the hands of today's younger generations who will pass the torch to future generations. We have mapped the road to sustainable development; it will be for all of us to ensure that the journey is successful and its gains irreversible.
>
> (United Nations, 2015, A/RES/70/1: 13)

References

Beck, Ulrich (2006) *Cosmopolitan Vision*, Cambridge, MA: Polity Press.

Beck, Ulrich (2015) 'Emancipatory catastrophism: What does it mean to climate change and risk society?' *Current Sociology*, 63(1): 75–88.

Beck, Ulrich (2016) *The Metamorphosis of the World*, Cambridge, MA: Polity Press.

Beck, Ulrich & Levy, Daniel (2013) 'Cosmopolitanized nations: Re-imagining collectivity in world risk society', *Theory, Culture & Society*, 30(2): 3–31.

Beck, Ulrich & Sznaider, Natan (2006) 'Unpacking cosmopolitanism for the social sciences: A research agenda', *The British Journal of Sociology*, 57(1): 1–23.

Bohman, James, (2012) '*Jus post bellum* as a deliberative process: Transnationalizing peace-building', *Irish Journal of Sociology*, 20(2): 10–27.

Castoriadis, Cornelius (1987) *The Imaginary Institution of Society*, Cambridge MA: MIT Press.

Chakrabarty, Dipesh (2011) 'The future of the human sciences in the age of humans', *European Journal of Social Theory*, 20(1): 39–43.

Chernilo, D. (2017) 'The question of the human in the Anthropocene debate', *European Journal of Social Theory*, 20(1): 44–60.

EarthJustice (2017) 'Clean energy', available at https://earthjustice.org/climate-and-energy/clean-energy (accessed: 16 December 2017).

Fine, Robert (2006) 'Cosmopolitanism and violence: Difficulties of judgment', *The British Journal of Sociology*, 51(1): 49–67.

Fine, Robert (2009) 'Cosmopolitanism and human rights: Radicalism in a global age', *Metaphilosophy*, 40(1): 8–23.

Friends of the Earth Europe (2009) 'Overconsumption? Our use of the world's natural resources', available at www.foei.org/resources/publications/publications-by-subject/economic-justice-resisting-neoliberalism-publications/overconsumption-our-use-of-the-worlds-natural-resources (accessed 9 December 2018).

Greenpeace (2016) Arctic still threatened by oil drilling, Atlantic coast spared', available at www.greenpeace.org/usa/arctic-still-threatened-oil-drilling-atlantic-coast-spared/ (accessed: 9 December 2018).

Greppi, Edoardo (1999) 'The evolution of individual criminal responsibility under international law', *International Review of the Red Cross*, 835, available at www.icrc.org/eng/resources/documents/article/other/57jq2x.htm (accessed: 20 July 2018).

Habermas, Jurgen (1992) *Between Facts and Norms: Contributions to a Discourse Theory of Law and Democracy*, Cambridge, MA: MIT Press.

Held, David (1995) *Democracy and the Global Order: From the Modern State to Cosmopolitan Governance*, Cambridge: Polity Press.

Honneth, Axel (2007) *Disrespect: The Normative Foundations of Critical Theory*, Cambridge, MA: Polity Press.

Intergovernmental Panel on Climate Change (2001) Third Assessment Report, available at www.ipcc.ch/working-group/wg1/?idp=0 (accessed: 9 December 2018).

Intergovernmental Panel on Climate Change (2014) 'Synthesis report: Summary for policy makers' available at http://ipcc.ch/pdf/assessment-report/ar5/syr/AR5_SYR_FINAL_SPM.pdf (accessed: 20 December 2017).

International Energy Agency (2017) 'Fossil fuels' available at www.iea.org/tcp/fossilfuels/ (accessed: 18 December 2017).

Lancet Commission on Pollution and Health (2017) available at www.lancet.com/commissions/pollution-and-health (accessed: 18 December 2017).

McGlade, Christophe & Ekins, Paul (2015) 'The geographical distribution of fossil fuels unused when limiting global warming to 2°C', *Nature*, 517: 187–190.

Mehta, Sailesh & Merz, Prisca (2015) 'Ecocide – a new crime against peace?' *Environmental Law Review*, 17(1): 3–7.

Merz, Prisca, Cabanes, Valerie & Gaillard, Emillie (2014) 'Ending Ecocide – the next necessary step in international law', paper prepared for the 18th Congress of the International Association of Democratic Lawyers, Brussels, 15–19 April 2014.

NASA (2017) Global Climate Change: Vital Signs of the Planet, available at http://climate.nasa.gov/causes (accessed: 20 December 2017).

OECD (2004) Poverty and Climate Change: Reducing the Vulnerability of the Poor through Adaptation' available at www.oecd.org/env/cc/2502872.pdf (accessed: 20 July 2018).

Raupach, Michael R. & Canadell, Josep G. (2010) 'Carbon and the Anthropocene', *Current Opinion in Environmental Sustainability*, 2: 210–218.

Rawls, John (1993) *Political Realism*, New York: Columbia University Press.

Rawls, John (2001) *Justice as Fairness: A Restatement*, ed. Erin I. Kelly, Cambridge, MA: Harvard University Press.

Raz, Joseph (1986) *The Morality of Freedom*, Oxford: Clarendon Press.

Rose, Michael (2016) 'Constitutions, democratic self-determination and the institutional empowerment of future generations', *Intergenerational Justice Review*, 2: 56–69.

Sassen, Saskia (2003) 'Globalization or denationalization?' *Review of International Political Economy*, 10(1): 1–22.

Skillington, Tracey (2016) *Climate Justice and Human Rights*, New York: Palgrave Macmillan.

Thompson, Dennis. F. (2010) 'Representing future generations: Political Presentism and democratic trusteeship', *Critical Review of International Social and Political Philosophy*, 13(1): 17–37.

Touraine, Alain (1996) 'A sociology of the subject', in Clark, Jon & Diani, Marco (eds.) *Alain Touraine*, London: Falmer.

Touraine, Alain (1997) *What Is Democracy?* Boulder, CO: Westview Press.

UNICEF (2010) 'A brighter tomorrow: Climate change, child rights and intergenerational justice', UK Committee for UNICEF, available at https://www.unicef.org.uk/wp-content/uploads/2010/09/intergenerationaljustice.pdf (accessed: 15 December 2017).

UNICEF (2017) 'Written submission to the United Nations office of the High Commissioner for Human Rights on climate change and the full and effective enjoyment of the rights of the child' (6 January) available at www.ohchr.org/Documents/Issues/ClimateChange/RightsChild/UNICEF.docx (accessed: 15 December 2017).

Union of Concerned Scientists (2016) 'The hidden costs of fossil fuels' available at https://www.ucsusa.org/clean-energy/coal-and-other-fossil-fuels/hidden-cost-of-fossils (accessed: 16 December 2017).

United Nations (2013) 'Report of the UN Secretary General, 'http://www.academia. edu/9203050/UN_Report_on_Intergenerational_Solidarity_and_the_Needs_of_ Future_Generations_Concerns_about_future_generations_make_their_way_to_the_ global_political_agenda' available at www.futurejustice.org/bolg/guest-contribution/un-report-on-intergenerational-solidarity-and-the-needs-of-future-generations (accessed: 15 December 2017).

United Nations (2015) 'Transforming our world: The 2030 Agenda for sustainable development' (A/RES/70/1) available at https://sustainabledevelopment.un.org/post2015/ transformingourworld (accessed: 15 December 2017).

United Nations Convention Concerning the Protection of the World Cultural and Natural Heritage (1972) available at legal.un.org/avl/ha/ccpwcnh/ccpwcnh.html (accessed 9 December 2018).

United Nations General Assembly (2012) 'Resolution adopted on 27 July 66/288 "The future we want"' available at https://sustainable development.un.org/futurewewant.html (accessed: 18 December 2017).

Urry, John (2010) 'Consuming the planet to excess', *Theory Culture & Society*, 27(2–3): 191–212.

Urry, John (2016) *What Is the Future?* Cambridge, MA: Polity Press.

World Bank (2017) 'The 2017 Atlas of sustainable development goals: A new visual guide to data and development' (4 April) available at https://blogs.worldbank.org/ opendata/2017-atlas-sustainable-development-goals-new-visual-guide-data-and-devel opment (accessed 7 December 2018).

World Future Council (2013) 'Crimes against future generations' available at www.world futurecouncil.org/crimes-against-future-generations/ (accessed: 20 July 2018).

World Future Council (2015) 'How can today's Europe better safeguard the needs of the future?' available at www.un.org/sustainabldevlopment/development-agenda (accessed: 15 December 2017).

World Health Organization (2014) '7 million premature deaths annually linked to air pollution' available at www.who.int/mediacentre/news/releases/2014/air-pollution/en/ (accessed: 16 December 2017).

Young, Iris Marion (2011) *Responsibility for Justice*, Oxford: Oxford University Press.

Index